JN223595

－動画で見る蚊の不思議な生態－

琉球列島の蚊

QR
コードから
23本の
蚊の動画が
見られる！

Mosquitoes in the Ryukyu Archipelago: Amazing biology of mosquito on video

23 mosquito videos can be viewed by scanning the QR code!

宮城一郎・當間孝子・岡澤孝雄・水田英生・比嘉由紀子著

Ichiro Miyagi, Takako Toma, Takao Okazawa, Hideo Mizuta and Yukiko Higa

動画一覧（略題） /Video list (Abbreviation titles)

●動画1J/Video 1E
・ネッタイイエカの生活史
・Life history of *Cx. quinquefasciatus*

●動画 2J/Video 2E
・ヒトスジシマカの生活史
・An aggressive biter *Ae. albopictus*

●動画 3J/Video 3E
・シナハマダラカの幼虫の摂食法
・Feeding habit of *An. sinensis* larva

●動画 4J/ 動画 4J
・オキナワヤブカ幼虫の摂食法
・Feeding mode of *Ae.o.okinawanus*

●動画 5J/Video 5E
・塩水に多発するトウゴウヤブカ
・Tidal brackish water pool breeder, *Ae. togoi*

●動画 6J/Video 6E
・アオミドロを摂食するカラツイエカの幼虫
・*Cx. bitaeniorhynchus* larvae, habitual shredder of *Spirogyra*

●動画 7J/Video 7E
・ミナミハマダライエカ幼虫の摂食方法
・*Cx. mimeticus*, employing two feeding modes

●動画 8J/Video 8E
・ミナミハマダライエカとカラツイエカ幼虫の摂食方法の比較
・Feeding habits of *Cx.mimeticus* and *Cx. bitaeniorhynchus* larvae

●動画 9J/Video 9E
・湿地性植物の根に付着するアシマダラヌマカの幼虫
・*Ma.uniformis* larva attached to water plant

●動画 10J/Video 10E
・汚水溜まりのお掃除屋さん，オオクロヤブカ幼虫
・Decomposer, *Ar. subalbatus* larva

●動画 11J/Video 11E
・ヤンバルギンモンカ I. 生息環境
・*To. yanbarensis* 1. Habitat

●動画 12J/Video 12E
・ヤンバルギンモンカ II. 交尾行動
・*To. yanbarensis* 2. Mating habit

●動画 13J/Video 13E
・ヤンバルギンモンカ III. 産卵行動
・*To. yanbarensis* 3. Oviposition behavior

●動画 14J/Video 14E
・ヤンバルギンモンカ IV. 幼虫の摂食行動
・*To. yanbarensis* 4. Feeding behavior of larva

●動画 15J/Video 15E
・トラフカクイカ幼虫の捕食
・Predators, *Lt. vorax* larvae

●動画 16J/Video 16E
・ユスリカの幼虫を捕食するヤエヤマオオカ
・*Tx. m. yaeyamae* preying on *Chironomid* larvae

●動画 17J/Video 17E
・人の血を吸うオオハマハマダラカ
・*An. saperoi* feeding on human

●動画 18J/Video 18E
・クロフトオヤブカの交尾行動
・Mating behavior of *Ve. iriomotensis*

●動画 19J/Video 19E
・魚の血を吸うカニアナヤブカ
・*Ae. baisasi* feeding on the mudskipper

●動画 20J/Video 20E
・カエルの鳴き声に誘引されるチスイケヨソイカとチビカ
・*Corethrella* and *Uranotaenia* attracted to frog calls

●動画 21J/Video 21E
・蟻の口から栄養源を頂戴するカギカの生態
・Feeding habits of *Malaya leei*

●動画 22J/Video 22E
・蚊の吸血嗜好性
・Foods preference of mosquitoes

●動画 23J/Video 23E
・蚊幼虫の基本的な摂食法
・Feeding modes commonly used by mosquito larvae

PDF　初心者のための日本産蚊科幼虫の検索図

動画で見る蚊の不思議名な生態—序にかえて

蚊と言えば人の血を吸い，マラリアやデング熱などの伝染病を媒介する嫌なことばかりをイメージするだろう．蚊は主として医学，公衆衛生学分野の研究対象となり，病原媒介に関係する限られた蚊の成虫の生態や分類の研究が古くから盛んに行われてきた(栗原, 2013)．双翅目（ハエ目）昆虫の中では蚊科（Culicidae）は最もよく研究されたグループと言える（Tanaka et al., 1979；田中，2014; 宮城・當間，2017；上村，2016，2017；津田，2019）．しかし，現在我が国に生息する124種の蚊の多くは病原媒介とは関係がなく野山に生息し，野生動物の血を吸って繁殖している．全く血を吸わない蚊もおり，生態がまだ良くわからない種も多い．蚊は生涯を通して生息する環境範囲，ニッチ（niche）は陸圏と水圏で，多種多様の特徴は成虫だけでなく，すべての発育期（蛹，幼虫）の形態，生態（行動）にも見られる．

蚊の幼虫・ボウフラは何をどのようにして食べているの？ 血を吸わない蚊ってどんな蚊？ 蚊の雄は血を吸わないと聞いたが，何を食べているの？ 蚊の祖先は？など，身近な蚊に関する疑問は多い．これらの質問に対する回答は近年発行されている蚊に関する参考書（Clements，1992；池庄司，1993；茂木，1999；宮城，2002；三條場ら，2019；一盛，2021；Wilkerson et al，2021）や最近ではネットで簡単に調べることができる．しかし，その回答を一読し，写真を見ただけでは実態は理解しにくい．著者（宮城）も蚊の研究を始めた当時，国内外の著名な蚊の文献やテキストにより蚊を同定し，ボウフラの摂食行動や成虫の交尾行動の記述に繰り返し目を通し，すでに理解したと思い込んでいた．しかし，読み違いが多く，撮影した動画を繰り返し見ることによりはじめて実態がよく見え，上記の文献やテキストに記述されている行動が理解できた．

最近の動画撮影機具や技術，情報公開の手段の発展は目覚ましい．希にみる蚊の行動を手持ちのデジカメやスマートフォンで撮影した動画を拡大して繰り返し見ることができる．動画サイトにアップした映像を全世界の人が見ることができるようになった．また，最近開発された動画編集ソフトを駆使し，人工知能（AI）で音読させ，さらに背景音楽（BGM）を使って編集することも容易になった．私達が研究対象にしている蚊の生態・行動に関しても最新のデジタル化の技術を取り入れ，これまでの情報を再検討し，新しい情報を発信することが望ましい．動画の記録は蚊の環境適応や進化学的な研究にも重要である（砂原・比嘉，2016)．

Mosquito behavior on video – an introduction

When we think of mosquitoes, we tend to think of unpleasant insects, such as sucking human blood and transmitting infectious diseases such as malaria and dengue fever. Mosquitoes have been the subject of research primarily in the fields of medicine and public health. For this reason, research into the biology and classification of the vector mosquitoes has been actively conducted for a long time (Kurihara, 2013). Among the Diptera, the Culicidae is the most well-studied group (Tanaka et al., 1979; Miyagi and Toma, 2017; Kamimura, 2017; Tsuda, 2019). However, most of the 124 species of mosquitoes currently recorded in Japan are not directly related to disease transmission and they live in the mountains and remote areas, breeding by sucking the blood of wild animals. There are also mosquitoes that do not suck blood at all, and the biology of these species is still not well understood. Mosquitoes inhabit both terrestrial and aquatic environments throughout their lives, and their diverse morphology and behavior can be seen not only in the adults but also in all stages of development (pupae and larvae).

What do mosquito larvae eat and how do they eat them? What kinds of mosquitoes don't suck blood? We know that male mosquitoes don't suck blood, but what do they eat? What are the ancestors of mosquitoes? There are many questions about the mosquitoes that we are familiar with. The answers to these questions can be easily found in recently published reference books on mosquitoes (Clements, 1992; Ikeshoji, 1993; Mogi, 1999; Miyagi, 2002; Sanjoba et al., 2019; Ichimori, 2021; Wilkerson et al., 2021) or online. However, it is difficult to understand the actual situation just by reading the answers to these questions and looking at fragmentary photographs. When the author (Miyagi) first began his research on mosquitoes, he identified them through well-known mosquito literatures and textbooks from Japan and abroad, and repeatedly read descriptions of the feeding behavior of mosquito larvae and the mating behavior of adults, believing that he already understood them. However, there were many misinterpretations, and it was only by watching the video footage repeatedly that he was able to get a clear picture of the situation and understand the behavior described in the literatures and texts mentioned above.

Recent developments in video-taking equipment and technology, as well as the means of disclosing information, have been remarkable. Videos on rare mosquito behavior captured by a digital camera or smartphone can be enlarged and viewed repeatedly in real time. It is now possible to instantly transmit videos to the world. In addition, by making full use of recently developed video editing software, it has become easy to have artificial intelligence (AI) read aloud and even edit using background music. It is desirable to incorporate the latest digitalization technology into the ecology and behavior of mosquitoes, which are the subject of our research, to reexamine existing information and disseminate new information. Video recording is also important for research into environmental adaptation and evolution of mosquitoes (Sunahara and Higa, 2016).

動画撮影，編集に際して留意した点

1. 対象とする蚊の分類・生態に熟知していること．
   今回は主として琉球列島に生息する蚊を対象にした．
2. 撮影は可能な限り自然の状態で行うこと．
   幼虫はプラスチックあるいは，ガラス製の水槽（4.8 × 5.5 × 2 cm）を用い，幼虫が生息していた溜まり水や 堆積物を入れて可能な限り自然光下で撮影した．成虫は観察箱（53 頁参照）を用いた．撮影機は運台付 3 脚を用い，スマートフォン，ニコンデジタルカメラ（D5600），オリンパスデジタルカメラ（タフ），ニコン顕微鏡を用いた．
3. 撮影した膨大な動画から各種の生態的特徴を抜粋し，編集した．編集用アプリ（Power Director），人工音声，無料背景音楽を使用した．

Points to keep in mind when filming and editing the video

1. Be familiar with the classification and ecology of the mosquitoes you are targeting. In our endeavor, we focused mainly on mosquitoes that live in the Ryukyu Archipelgo.
2. Photographs should be taken in natural conditions as much as possible. Larvae were photographed in plastic or glass tanks (4.8×5.5×2 cm) filled with the standing water or sediment in which the larvae were collected, and photographed under natural light as much as possible. Adults were observed using an observation box (see page 53). The photographs were taken on a tripod with a stand, using a smartphone, a Nikon digital camera (D5600), an Olympus digital camera (Tough), and a Nikon microscope.
3. Various ecological characteristics were extracted from the vast amount of footage and edited. An editing application Power Director, artificial voices (AI), and free background music were used.

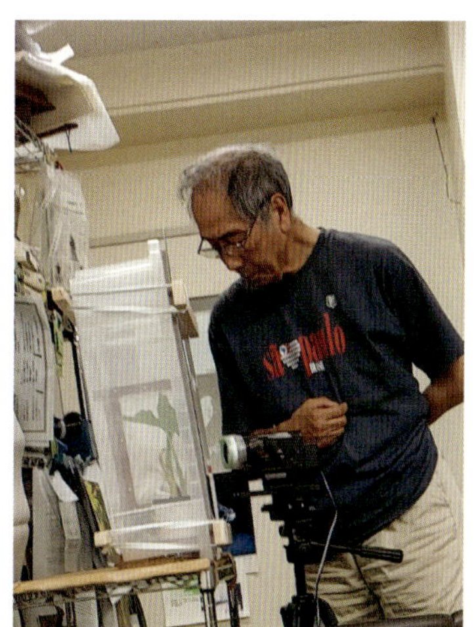

水槽の中の幼虫の撮影と観察箱内の成虫を観察
Photographing mosquito larvae in an aquarium and observing adults in the box

この本の見方

本書は2017年に宮城・當間が東海大学版部から出版した『琉球列島の蚊の自然史』の姉妹編と考えている．身近に見られる蚊の吸血行動や幼虫の遊泳，捕食行動を折に触れて撮影した動画や写真を編集ソフトと人工知能や背景音楽を用いて23動画に編集（日本・英語版）した．それらを動画サイトにアップし，そのサイトのQRコードをスマートフォンなどで読み取れば，動画を見ることができるようにした．

また本書には資料編として蚊の分類，生態や吸血嗜好性に関して付表1－4にまとめた．さらに，水田が検疫所内部資料として作成した初心者のための日本産蚊科幼虫の検索図(日本語版）を掲載し，PDF版をQRコードから読み取りができるようにした．

本書を手にされた読者はまず本書を一読し，その後動画を見ていただきたい．すべての動画はQRコードで見ることができるが，全て見ると2時間30分になる．時間がない方は興味のある動画から見ていただきたい．さらに動画を見た後にもう一度本書に目を通していただきたい．あるいは逆に本書を見返して，その後に興味のある動画を注視していただいてもよろしい．

How to read this book

This book is considered a sister volume to “The Natural History of Mosquitoes in the Ryukyu Archipelago” by Miyagi and Toma (2017), published by Tokai University Press. Videos and photos of mosquitoes’ blood-sucking, larval swimming, and feeding behaviors, which can be seen all around us, were taken from time to time both inside and outside the laboratory. Using editing software, artificial intelligence, and background music, we edited videos and photos taken from time to time of the blood-sucking behavior, larval swimming, and predatory behavior of mosquitoes that we see all around us into 23 videos (Japanese and English versions). Explanations have been added, and the videos can be viewed by scanning the QR code with a smartphone or other device. Important information on mosquito classification, ecology, and blood-sucking preferences has been compiled in Appendices 1－4. In addition, a new illustrated key table (Japanese version only) of Japanese mosquito larvae for beginners, which Mizuta created as internal documentation for the quarantine station, has been added and can be read using the QR code. We would like readers of this book to first read the concluding remarks in the main text and then watch the videos. All videos can be viewed using QR codes, and it will take 2 hours and 30 minutes to watch them all. If you don’t have much time, you can start by watching a video that interests you. After watching the video, please read the conclusion of the main text again. Alternatively, you could read the conclusion of the main text first, and then watch the video that interests you.

琉球列島の蚊
－動画で見る蚊の不思議な生態－

Mosquitoes in the Ryukyu Archipelago:
Amazing biology of mosquitoes on video

目次
Contents

第１章 動画を見る前に知っておきたい蚊の基本的な形態および生態
Chapter 1. Basic information about mosquito morphology and biology that you should know before watching the video

第２章　動画で見る蚊の不思議な生態
Chapter 2. Amazing biology of mosquitoes on video

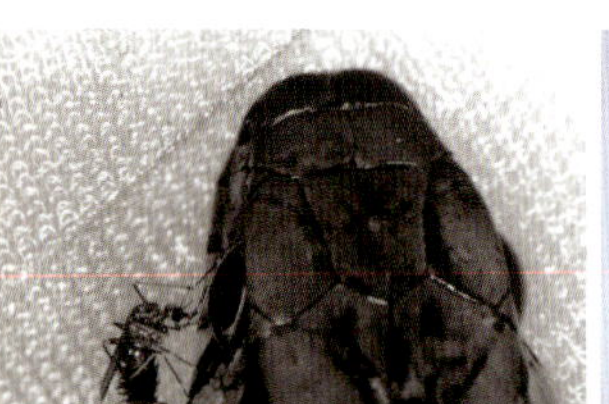

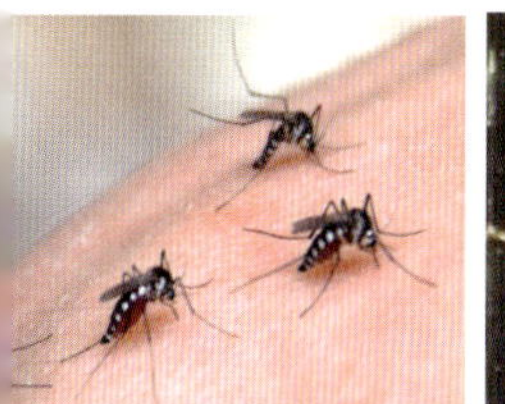

第3章 初心者のための日本産蚊科幼虫の検索図（主として4齢幼虫）

付表 Appendix

がじゃんコラム

【蚊帳（かや）とのかかわり】

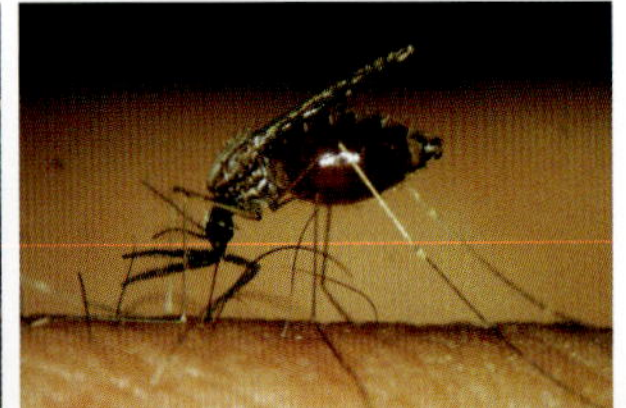

第１章
動画を見る前に知っておきたい蚊の基本的な形態および生態

Chapter 1.
Basic information about mosquito morphology and biology that you should know before watching the video

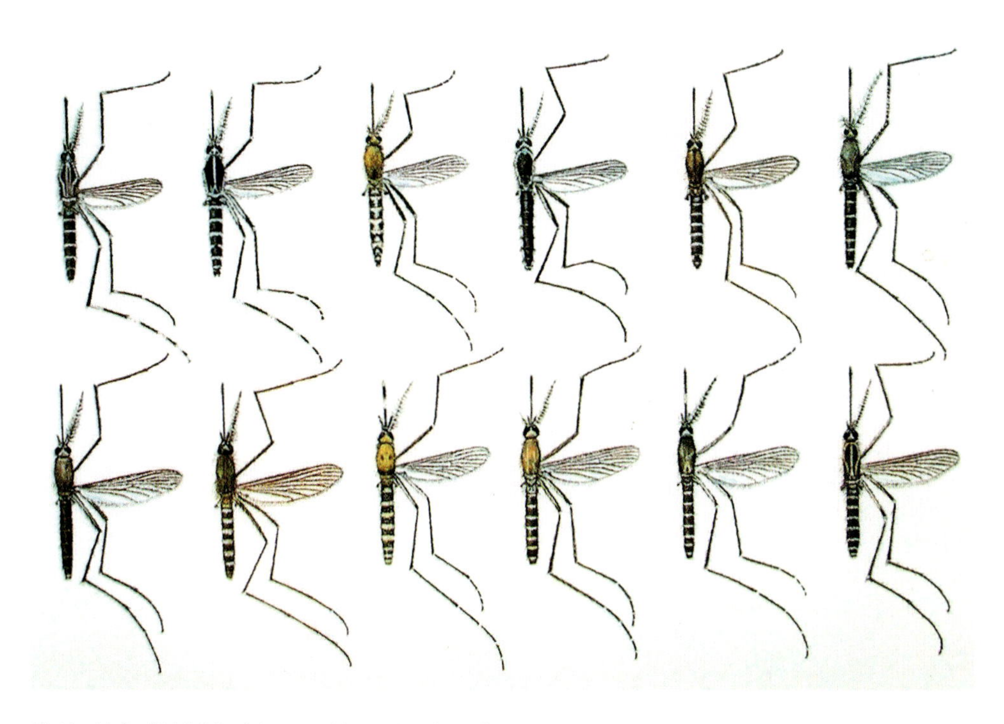

琉球列島産蚊種（太田和鐘三氏　描写）
Some mosquitoes collected from the Ryukyu Island (painted by Mr. S. Ohtawa)

琉球列島の蚊相

現在，日本全体の蚊相は亜種を種として数えると 124 種からなる（田中，2014）．琉球列島からは 77 種が記録されている（付表 1，2）．琉球列島の面積（3,589 km^2）は日本全土の面積（374,054 km^2）の 1/100 にすぎないが，他の地域の蚊相に比較して遥かに多種多様である．台湾，南中国，フィリピンなどの東洋区に共通して分布する種は 49（63.6%），九州以北の旧北区日本に共通して分布する種は 28（36.4%），東洋区と旧北区日本に共通して分布する種は 22（28.6%），固有種は 22 種（28.6%）である．本書においては主として沖縄島と西表島で採集・撮影した主要な蚊 16 種とチスイケヨソイカについての行動を 23 動画に収録した .

Mosquito fauna of the Ryukyu Archipelago

Currently, the mosquito fauna in Japan consists of 124 species, including subspecies (Tanaka, 2014). Of these, 77 species have been recorded from the Ryukyu Archipelago (Appendices 1, 2). Although the Ryukyu Archipelago (3,589 km^2) is only 1/100 of the total area of Japan (374,054 km^2), its mosquito fauna is far more diverse than those of other regions. There are 49 species (63.6%) that are distributed in the Oriental Region, including Taiwan, southern China, and the Philippines; 28 species (36.4%) that are distributed in the Palaearctic region of Japan north of Kyushu; 22 species (28.6%) that are distributed in both the Oriental Region and Palaearctic Japan; and 22 species (28.6%) that are endemic. This book contains information on the behaviors of several species that were collected and photographed mainly on Okinawajima and Iriomotejima.

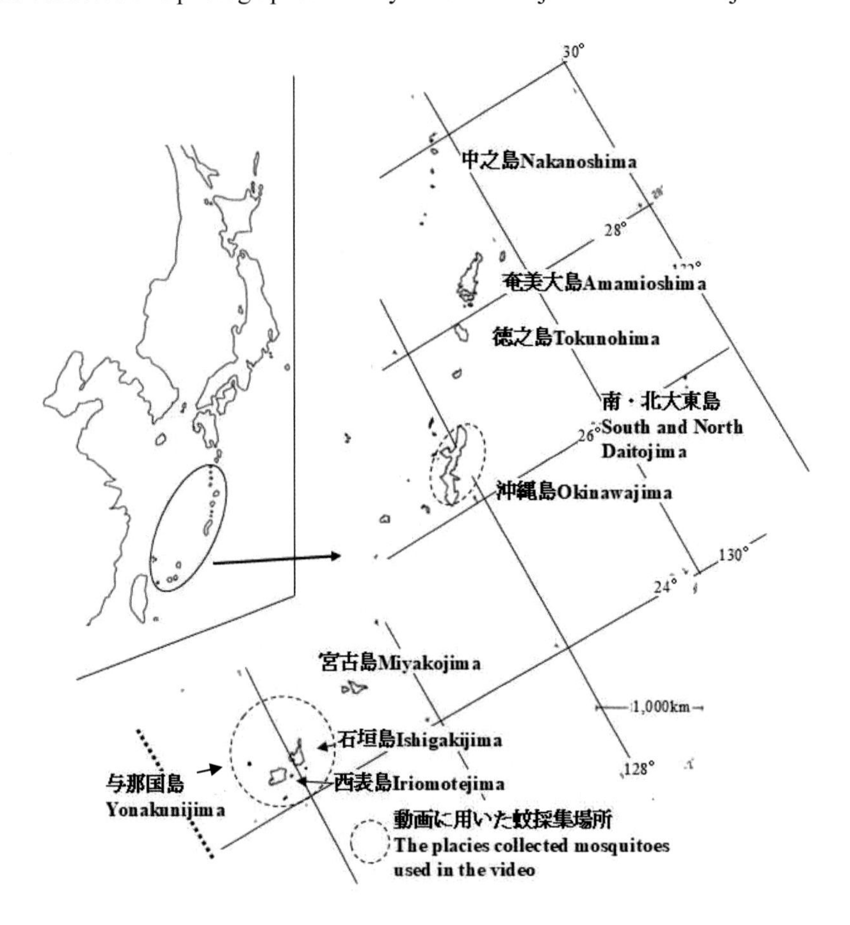

琉球列島で記録がある蚊 77 種と，動画に登場する蚊 16 種の種名・学名とチスイケヨソイカ

Japanese and scientific names of 77 mosquito species recored from the Ryukyu Archipelago, and 16 mosquito species and one *Corethrella* sp. that appear in the video.

蚊 No.		動画 No.	蚊 No.		動画 No.	蚊 No.		動画 No.	蚊 No.		動画 No.
1	モンナシハマダラカ *An. bengalensis*		21	ダイトウシマカ *Ae. daitensis*		41	クロフクシヒゲカ *Cx. nigropunctatus*		61	アシマダラヌマカ *Ma. uniformis*	⑨
2	オオツルハマダラカ *An. lesteri*		22	ダウンスシマカ *Ae. f. downsi*		42	アカクシヒゲカ *Cx. pallidothorax*		62	ハマダラナガスネカ *Or. anopheloides*	
3	ヤマトハマダラカ *An. l. japonicus*		23	ミヤラシマカ *Ae. f. miyarai*		43	リュウキュウクシヒゲカ *Cx. ryukyensis*		63	オキナワカギカ *Ml. genurostris*	㉑
4	オオハマハマダラカ *An. saperoi*	⑰	24	リバースシマカ *Ae. riversi*		44	カギヒゲクロウスカ *Cx. brevipalpis*		64	ヤンバルギンモンカ *To. yanbarensis*	⑪ ⑫ ⑬ ⑭
5	シナハマダラカ *An. sineisis*	③	25	トウゴウヤブカ *Ae. togoi*	⑤	45	リュウキュウクロウスカ *Cx. h. ryukyuanus*		65	ヤエヤマナガハシカ *Tp. yaeyamensis*	
6	タテンハマダラカ *An. tessellatus*		26	オオクロヤブカ *Ar. subalbatus*	⑩	46	オキナワクロウスカ *Cx. okinawae*		66	ヤエヤマオオカ *Tx. m. yaeyamae*	⑯
7	コガタハマダラカ *An. yaeyamaensis*		27	アマミムナゲカ *Hz. kana*		47	クロツノフサカ *Cx. bicornutus*		67	ヤマダオオカ *Tx. m. yamadai*	
8	キンイロヤブカ *Ae. v. nipponii*		28	コガタフトオヤブカ *Ve. nobukonis*		48	ハラオビツノフサカ *Cx. cinctellus*		68	オキナワオオカ *Tx. okinawensis*	
9	オキナワヤブカ *Ae. o. okinawanus*	④	29	アカフトオヤブカ *Ve. atriisimilis*		49	フトシマツノフサカ *Cx. infantulus*		69	カニアナチビカ *Ur. jacksoni*	
10	ヤエヤマヤブカ *Ae. o. taiwanus*		30	クロフトオヤブカ *Ve. iriomotensis*	⑱	50	アカツノフサカ *Cx. rubithoracis*		70	ムネシロチビカ *Ur. nivipleura*	
11	ニシカワヤブカ *Ae. nishikawai*		31	オビナシイエカ *Cx. fuscocephala*		51	カニアナツノフサカ *Cx. tuberis*		71	リュウキュウクロホシチビカ *Ur. n. ryukyuana*	
12	カニアナヤブカ *Ae. baisasi*	⑲	32	ジャクソンイエカ *Cx. jacksoni*		52	カラツイエカ *Cx. bitaeniorhynchus*	⑥	72	シロオビカニアナチビカ *Ur. ohamai*	
13	ムネシロヤブカ *Ae. albocinctus*		33	ミナミハマダライエカ *Cx. mimeticus*	⑦	53	ミツホシイエカ *Cx. sinensis*		73	イリオモテチビカ *Ur. tanakai*	
14	アマミヤブカ *Ae. j. amamiensis*		34	シロハシイエカ *Cx. pseudovishnui*		54	サキジロカクイカ *Lt. fuscana*		74	ハラグロカニアナチビカ *Ur. yaeyamana*	
15	サキシマヤブカ *Ae. j. yaeyamensis*		35	ネッタイイエカ *Cx. quinquefasciatus*	①	55	トラフカクイカ *Lt. vorax*	⑮	75	オキナワチビカ *Ur. annandalei*	
16	ナンヨウヤブカ *Ae. lineatopennis*		36	ヨツホシイエカ *Cx. sitiens*		56	オキナワエセコブハシカ *Fi. ichiromiyagii*		76	コガタチビカ *Ur. lateralis*	
17	ハマベヤブカ Ae. *vigilax*		37	コガタアカイエカ *Cx. tritaeniorhynchus*		57	マダラコブハシカ *Mi. elegans*		77	マクファレンチビカ *Ur. macfarlanei*	
18	ワタセヤブカ *Ae. watasei*		38	スジアシイエカ *Cx. vagans*		58	ルソンコブハシカ *Mi. luzonensis*			チスイケヨ *Corethrella* biting midges	⑳
19	ネッタイシマカ *Ae. aegypti*		39	ニセシロハシイエカ *Cx. vishnui*		59	ムラサキヌマカ *Cq. crassipes*				
20	ヒトスジシマカ *Ae. albopictus*		40	セシロイエカ *Cx. whitmorei*		60	キンイロヌマカ *Cq. ochracea*				

蚊の生活史

蚊は完全変態を行う昆虫で，一生のうちに卵・幼虫・蛹期は水中で，成虫期は陸上で生活をする．各発育期の生態は種により多少異なるが基本的には下の図ようである．卵（1）は粒状（ヤブカ属，ハマダラカ属）と卵塊（イエカ属）があり，卵期間は通常 2, 3 日．孵化した 1 齢幼虫は水中の有機物を摂食し，3 回脱皮して，4 齢幼虫（2），幼虫期間は 5, 6 日，4 齢幼虫は脱皮（蛹化）した蛹（3）は 2, 3 日間，全く摂食せず自由に遊泳する．羽化成虫（4）は 2 日後には交尾（5）を終え，雌は吸血（6）し，2 日後には産卵する．沖縄では蚊のことをガジャンと呼んでいる．ガジャンは冬眠することなく，年中繁殖している．

Life history of mosquitoes

Mosquitoes are insects that undergo complete metamorphosis, with the adult stage living on the land and the egg, larval and pupal stages living in the water. The biology of each developmental stage varies slightly depending on the species. There are 2 types (1) of egg laying habit, singly-laying (*Aedes* and *Anopheles*) and raft-laying (*Culex*). Egg stage period is 2 to 3 days. Larva (2) molts 4 times during period of 5 to 7 days. Larvae feed on small organic and inorganic particles. Pupal stage (3) period 2 to 3 days. The pupa is quite active. Adult emergence (4). Mating (5). Bood feeding (6). Two days after blood-feeding lay eggs. In Okinawa, mosquitoes are called "gajyan". They do not hibernate and breed all year round.

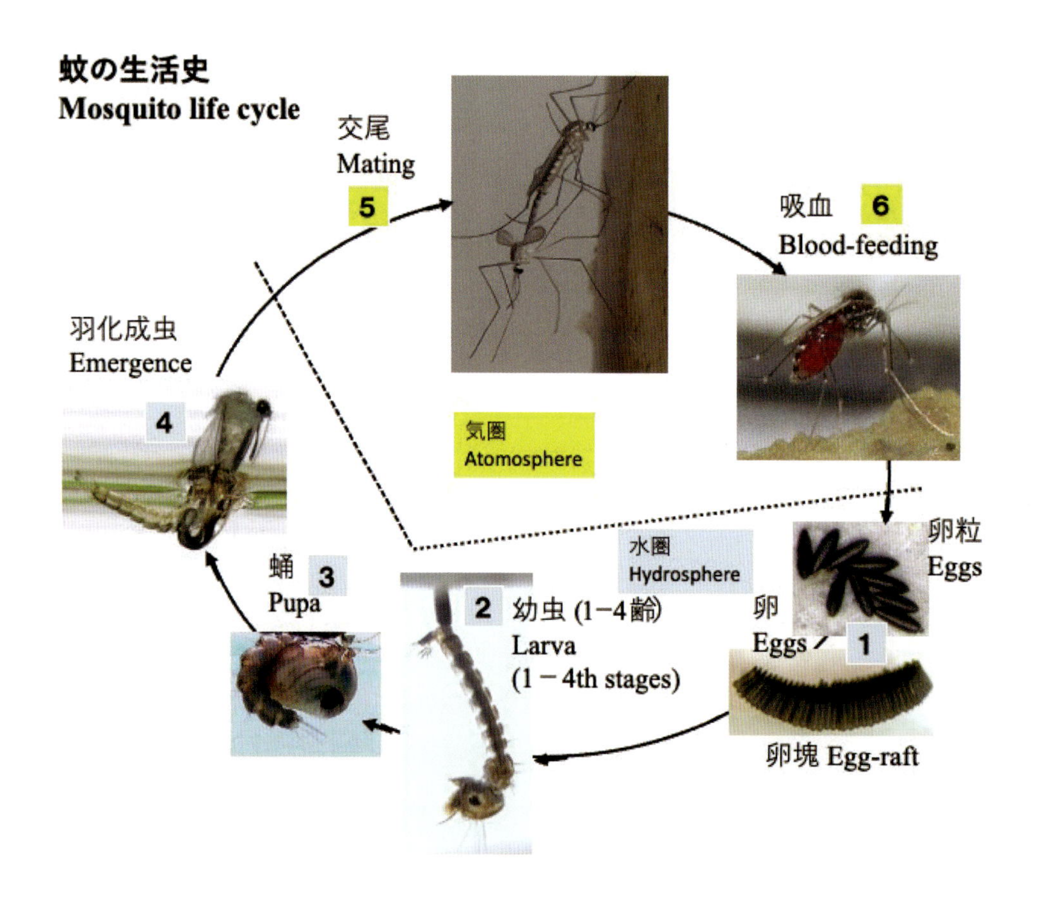

カ科（Culicidae）と姉妹科チスイケヨソイカ（Corethrellidae）の形態的特徴

2枚の翅をもつ昆虫はハエ目に分類されるが，蚊はか細い体に，胸部から長い3対の脚と頭部から前方に細長く伸びた口吻を有するので他のハエ類とは容易に区別できる．また口吻が長く伸びているハエ類には，蚊の他にツリアブ科（Bombyliidae）やオドリバエ科（Empididae），アブ科（Tabanidae）などの一部の種が含まれるが，翅脈（Wing vein）の脈相は蚊科と著しく異なる．蚊の祖先系と考えられているケヨソイカ科の翅脈相は蚊科の脈相とほとんど同じだが．翅の縁毛が著しく，吻は未発達で蚊科とは容易に区別される．比較表は次ページに掲げた。

Morphological characteristics of Culicidae and sister family Corethrellidae

Mosquitoes have a slender body, three pairs of long legs extending from the thorax, and a proboscis that extends forward from the head, making them easily distinguishable from other flies. Flies with long proboscis include, in addition to mosquitoes, some species of Bombyliidae, Empdidae, and Tabanidae, but the venation of their wing is significantly different from that of Culicidae. The venation of the Corethrellidae, which is thought to be the ancestor of mosquitoes, is almost the same as that of Culicidae, but the wing bristles are prominent and the proboscis is underdeveloped, making them easily distinguishable from Culicidae.

オオハマハマダラカの雌成虫
Anopheles saperoi

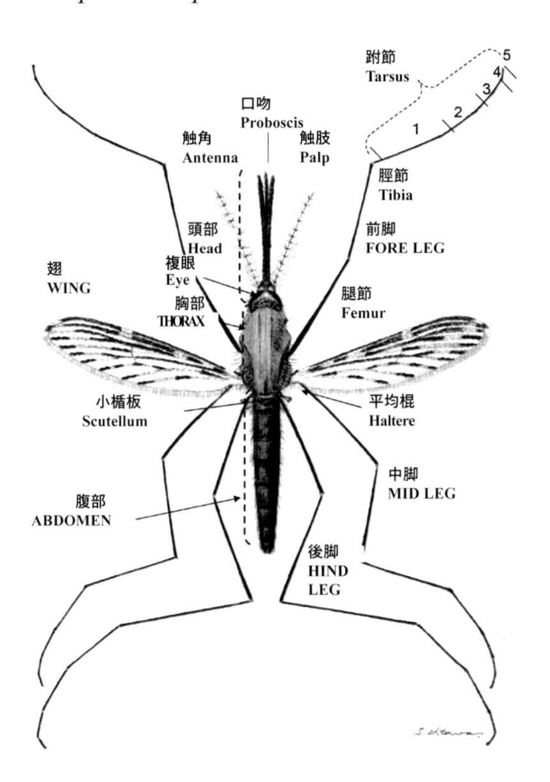

イエカの幼虫
Culicine larva

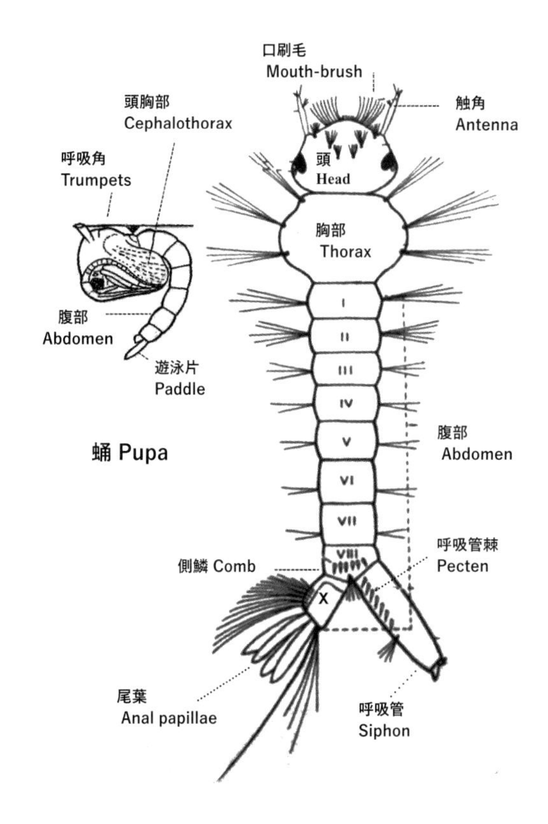

ハエ目各科の翅脈
Wing venations of different family in Diptera

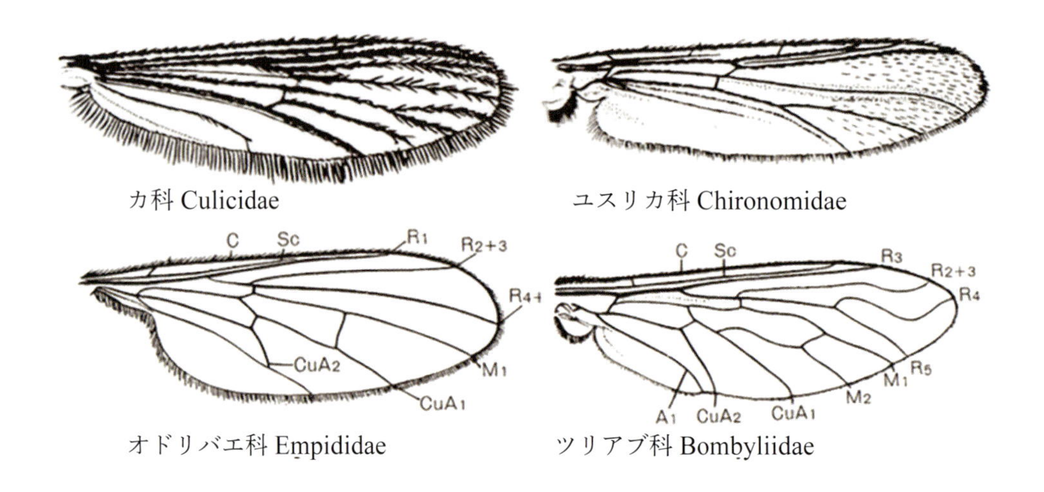

コガタアカイエカとニホンチスイケヨソイカの形態的相違
Differences in morphology between *Cx. tritaeniorhynchus* and *C. nippon*

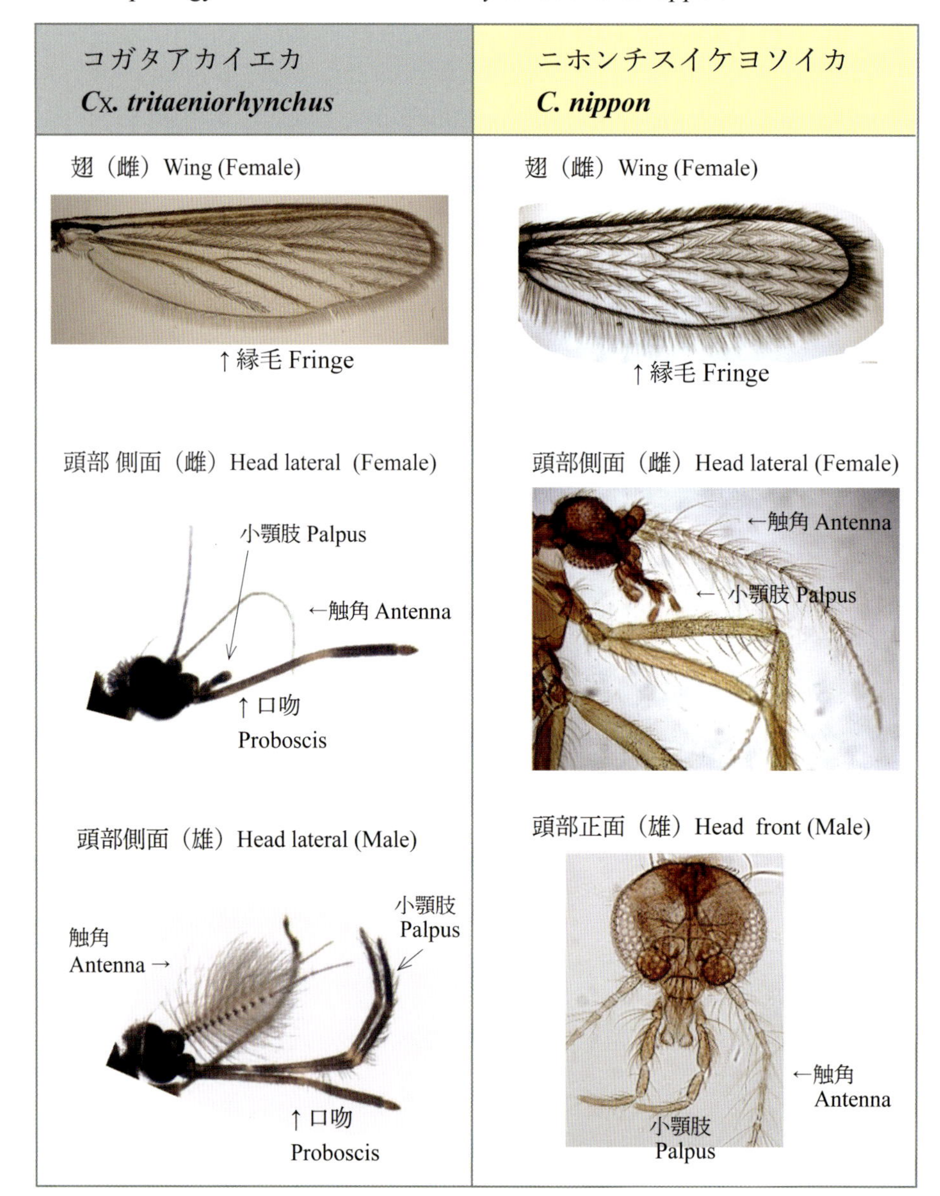

がじゃんコラム

【蚊帳(かや)とのかかわり】
その1．必需品としての蚊帳

私（昭和 11 生）は多感な青少年時代（昭和 20 年代）を奈良県朝和村（現在の天理市永原）で過ごした．数多くの思い出の中の一つに蚊帳とのかかわりがある．平成，令和生まれの若者には蚊帳を使用した人はもちろんのこと，蚊帳を見たこともない人が多い．

奈良大和盆地の夏は蒸し暑い．自宅の裏は水田が広がり，台所の排水とつながる溝の汚水溜まりは蚊の好適な発生源だった．家屋の窓には網戸はなく，勿論クーラーがない．我が家は両親と 6 人兄弟姉妹で，開けっぱなしの家屋の窓から日暮れとともに大量の蚊が押し寄せた．夕食後，家族全員は 8 畳の部屋に蚊帳を張って雑魚寝であった．蚊帳を吊るのは私と姉の役目だった．部屋の四角に蚊帳の紐をかけて裾が畳（床）から離れないように高さを調整し，きちんと張ること．蚊帳に入る時はしゃがみ，姿勢を低くして蚊帳の裾を叩いて素早く入ることなど，母親から教わった．蚊帳は緑色で麻や木綿製で，通気が良く，時々通過するそよ風，軒下の風鈴の心地よい音，田んぼからの蛙の合唱を聞きながら親子そろって深い眠りについた．

早朝，皆が起き，誰もいない蚊帳の四角には決まって血を腹いっぱい吸った数個体の蚊がおり，手のひらで叩き殺した．これらの蚊の種類は今となれば知る由もないが，当時の周囲の環境から判断して，夜間吸血性の水田に発生するコガタアカイエカ，シナハマダラカ，それに汚水溜まりから発生するアカイエカなどと思われる．当時，蚊帳は必需品でどこの家にも大小 2，3 張はあった．

奈良県は蚊帳生地の生産地で，昭和 30 年頃には織機 4,250 台，全国の 1.3%を占め，年間 22 億円を生産していたようだ．昭和が終わるころから農薬や殺虫剤散布による蚊の発生抑制や環境衛生の機運の高まりによる上下水道の整備，建築様式の和式から洋式への変化に伴うクーラーの普及により蚊帳の生産量は著しく減少した．現在，我が国では蚊帳を目にすることはなくなった．蚊帳の中での何気ない親子の雑談は懐かしく，ノスタルジックな思い出となった．（宮城）

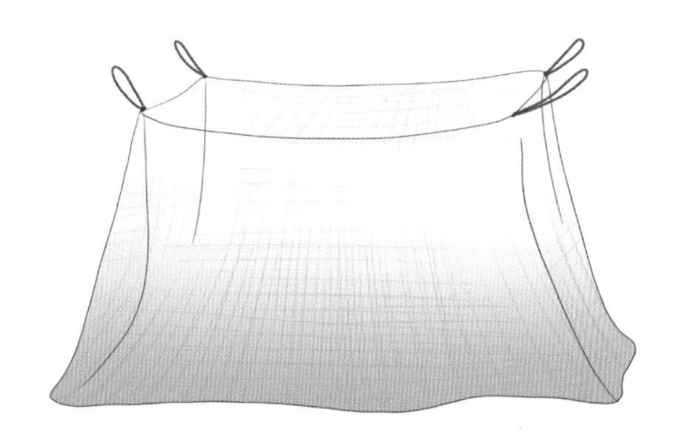

ヤンバルの多種多様な蚊の繁殖に関わる希少動物（村山望氏 撮影）
Rare animals involved in the reproduction of the Yanbaru's diverse mosquito fauna
（Photo by Mr. N. Murayama ）

第２章
動画で見る蚊の不思議な生態
chapter 2.
Amazing biology of mosquitoes on video

Collecting-filtering
泥中の筒巣の中のユスリカにかみついた

1. ネッタイイエカの生活史

動画 1J

上映時間
6 分

ネッタイイエカ *Culex quinquefasciatus* は熱帯地方に広く分布し，鹿児島以北の温帯地方に分布するアカイエカ *Culex pipiens* と亜種関係にあり，両亜種の生態・形態はよく似ている．沖縄ではこの蚊の幼虫はヒトスジシマカと共に人家周辺の人工容器の水溜まりに生息している．ネッタイイエカの成虫は，夜間，人を吸血し，かつて我が国で流行したフィラリア（バンクロフト糸状虫）の媒介蚊としてよく知られている．黒色の塵箱（プラスチック製）に汚水を入れて庭の片隅に放置すると，10 日後には決まってネッタイイエカとこの蚊の捕食性天敵トラフカクイカの幼虫が発生する．ネッタイイエカの幼虫は，体をくねらせ水中を後ずさり遊泳するほか，水面で体をくねらせることなく直進遊泳する．摂食は水面で呼吸管を上にして懸垂し，口刷毛を振動し，渦巻状の水流を起こして水中の浮遊物を口元に引き寄せ口腔内に取り込でいる．この摂食方法を収集濾過法と呼んでいる．

Life history of *Culex quinquefasciatus*

Video 1E

Show time
5 m 26 s

Culex quinquefasciatus is widely distributed in tropical countries. The adult is ready to feed on human blood in the house at night. It is an important vector of human filariasis. The larva is found commonly with *Aedes albopictus* in artificial containers around houses. Usually, movement of the larva is achieved by a side to side lashing of the whole body, which moves tail-first through the water. Another movement is gliding, in which the larvae propel themselves by means of the mouth-brushes. Feeding takes place near the water surface by filter-feeding. This is the typical mode of feeding among mosquito larvae. The feeding mode involves removal of particles suspended in the water current generated by the mouth-brushes. The current carries continuously newly suspended small particles towards the preoral cavity. The particles pass directly into the digestive tract.

卵塊 (5 – 8mm)，約 250 個の卵粒からなる

Egg raft (5 – 8mm) including about 250 eggs

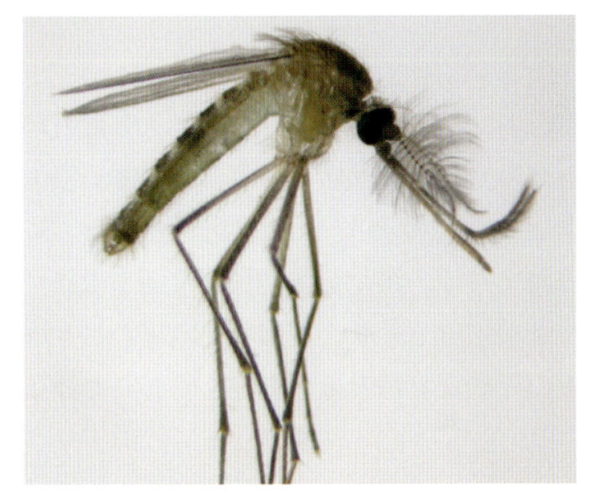

雄成虫　Adult male

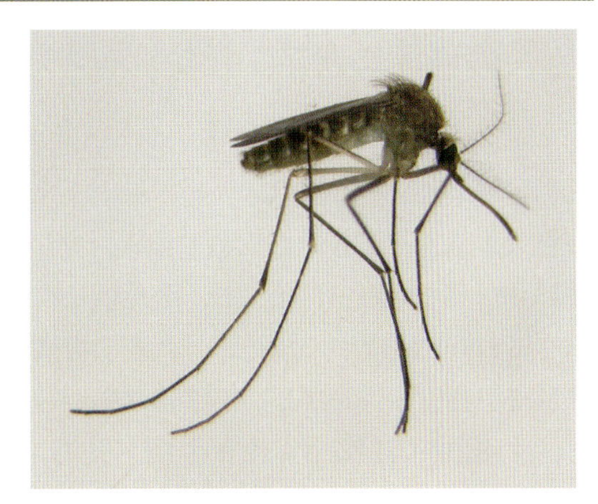

雌成虫　Adult female

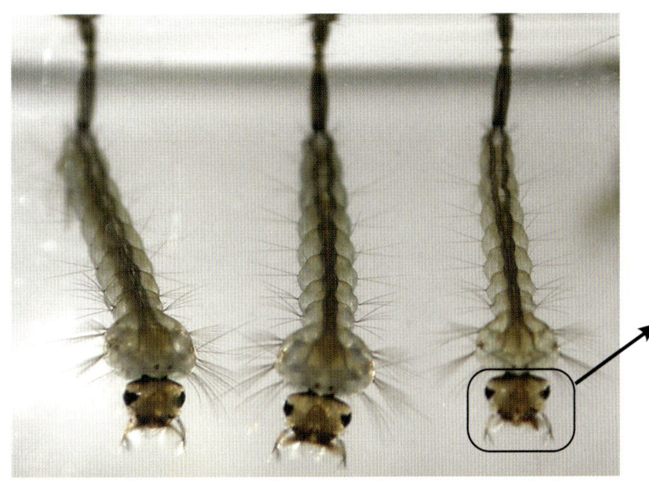

収集濾過法で水中の浮遊物を食べる幼虫
Feeding takes place near the water surface by filter-feeding

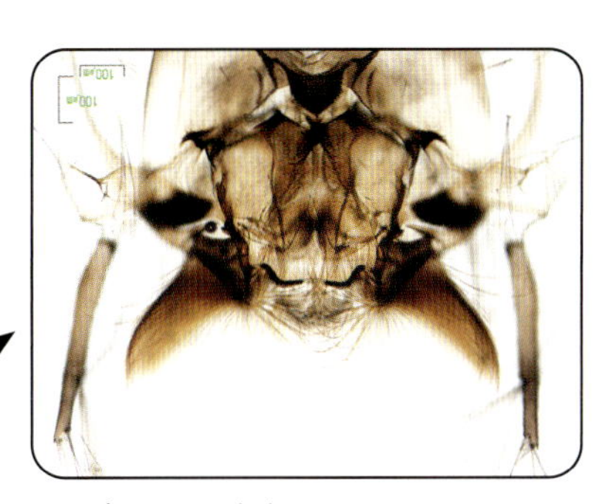

頭部の口刷毛
Mouth-brushes in head

蛹（鬼ボウフラ）Pupa

幼虫の主要な発生源，放置された人工容器
The main source of larvae is discarded man-made containers

2. 人の環境に適応したヒトスジシマカ

動画 2J

上映時間
5 分 08 秒

ヒトスジシマカはネッタイイエカと共に人家周辺に放置された空缶や古タイヤなどの人工容器に発生する蚊で，デング熱の媒介蚊としてよく知られている．昼間，木陰で吸血に飛来する．人工容器の内壁に産み付けられた卵は乾燥に耐えて 1 ヵ月間は生きている．沖縄では冬でも幼虫が発生し，暖かい日には人の血を吸う．孵化した幼虫は水面と水底で体をくねらせ後ずさり遊泳し，水底の堆積物を口刷毛でほじくり，有機物を収集かじり取り法で摂食する．しばしば堆積物の中から小昆虫の死骸など好みの有機物をくわえて浮上し，水面で呼吸しながら摂食する．

An aggressive day-time biter *Aedes albopictus*

Video 2E

Show time
4 m 32 s

*Aedes albopictu*s is a container breeder and well known as a vector of dengue fever. The adult female bites human easily during day-time in the shade. Their eggs, laid on the surface of the container, exhibit desiccation resistance, and can survive for a month without water. The larva swims by side to side lashing movement of the whole body, between bottom and water surface for feeding and respiration. The feeding mode of the larva is variable: collecting-gathering, scraping and shredding. The larva collects dead invertebrate in the bottom of the water, and comes the surface to shred it. Such feeding scenes are often observed in the aquarium.

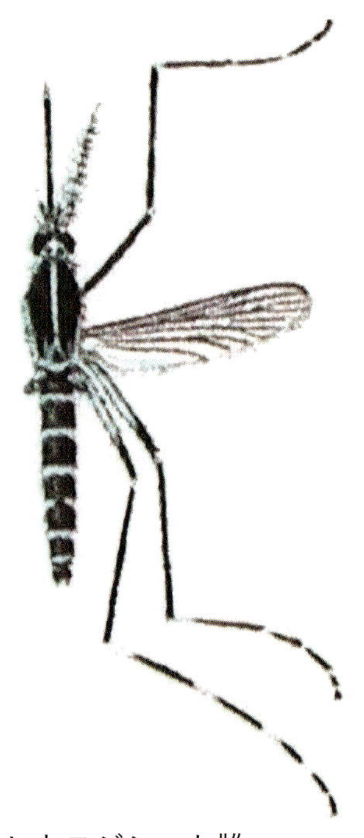

ヒトスジシマカ雌
Aedes albopictus female

ヒトスジシマカ幼虫
Aedes albopictus larva

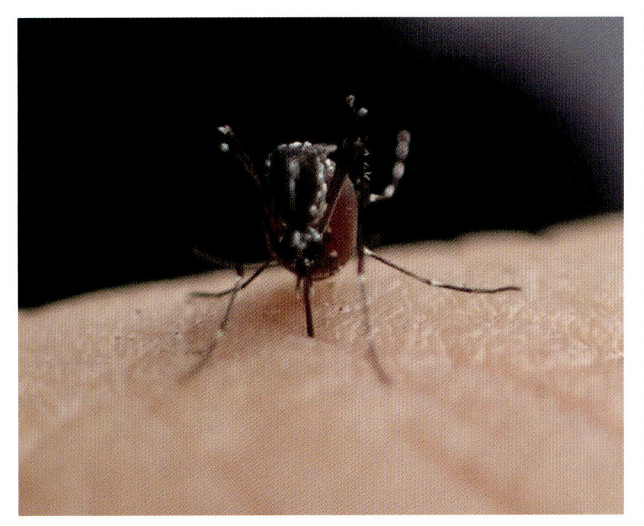

昼間木陰で人の血を吸うヒトスジシマカ
Ae. albopictus fed human in shed at day time

ヒトスジシマカの幼虫
Larva of *Ae. albopictus*

ゴミ捨て場の空缶などの水溜まりに好んで発生するヒトスジシマカの幼虫
The larvae preferentially grow in puddles such as empty cans in garbage dumps

空缶の内壁に産み付けられたヒトスジシマカの卵粒 乾燥に耐えて１カ月は卵殻内で生きている
The eggs laid on the inside walls of empty cans. They can survive for up to a month inside the egg shell, surviving the drying process.

3. 水表面で摂食するシナハマダラカの幼虫

動画 3J

上映時間
4 分 23 秒

ハマダラカ属の幼虫は呼吸管が未発達で常に水面に水平に静止して呼吸する．摂食は水表面で頭部だけを 180° 回転し，口器を上に向けて口刷毛を振動し，水流を起こし浮遊物を引き寄せ口元に取り込んで行う．引き寄せた浮遊物のうち微粒な有機物は咽頭内に直接取り込まれるが，大きな有機物は大顎，小顎と下唇基板でかみ砕いて咽頭内に取り込む．また，大きな固い物は頭部を 90° または 180° 素早く回転し側面や下方にはねのける．摂食は収集濾過法と時には収集かじり取り法で行なっている．水流には二つの方向があり，渦巻摂食と直流摂食に区別される．

Feeding habit of *Anopheles sinensis* larva

Video 3E

Show time
4 m 31 s

The siphon-less, *Anopheles* larvae are surface film feeders, feeding and breathing simultaneously at the water surface. The larvae have distinct methods of gathering foods by the collecting and filtering modes (Bate, 1949; Clements, 1992): “Eddy” and “Interfacial” feedings. The mouth-brushes produce a strong current of water in eddies from the area in front of the larva. In the “Interfacial” phase, there are no eddies and the floating foods approach the mouth along straight lines from all directions. The current is directed towards the mouth where the maxillae comb out large and solid particles and swallowed. This method is commonly observed. In order to get their foods floating on the water surface, all kinds of *Anopheles* larvae rotate their heads through 180 degrees, so that the side bearing the mouth-brushes becomes uppermost, to get easily at the foods. The unsuitable foods, large and solid particles, in the current are pushed aside by the head rotating through 90 or 180 degrees. The feeding mode is characteristic of the *Anopheles* larvae.

雌成虫　Adult female

翅の白斑　Wing spots of adult female

シナハマダラカの摂食に関係する口器
Larval mouthparts of *Anopheles sinenesis* associated with feeding habits

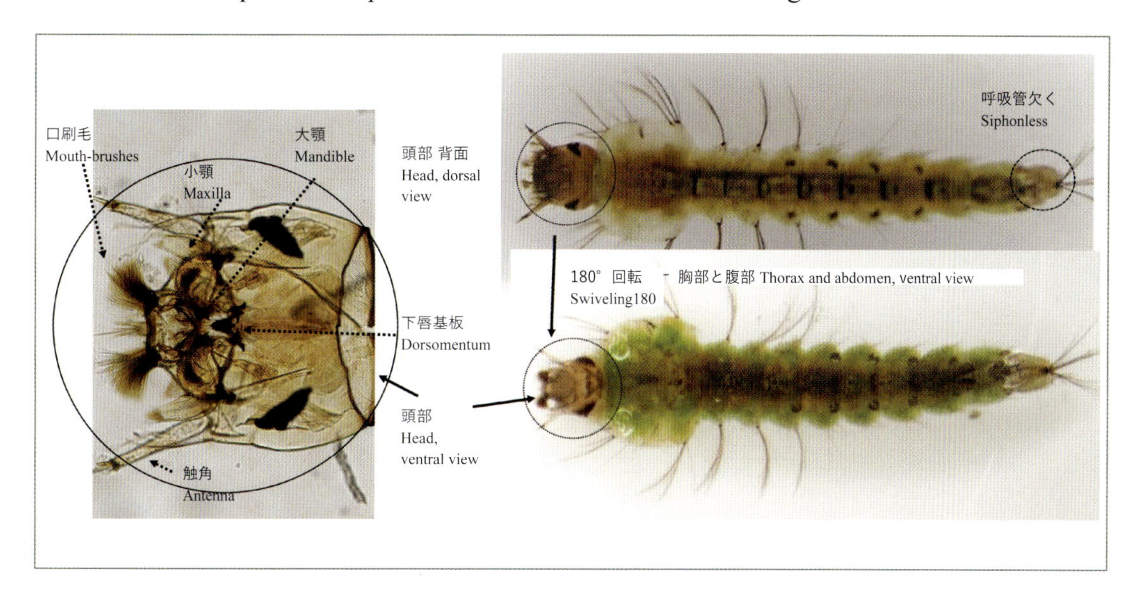

水面に水平に静止するハマダラカの幼虫（左）と懸垂静止するチビカの幼虫
Anopheles larvae resting horizontally on the water surface (left) and *Uranotaenia* larvae resting suspended from the surface

シナハマダラカ幼虫の発生源水田
Paddy fields as a breeding ground for *An. sinensis* larvae

ハマダラカ幼虫の摂食　収集濾過法のタイプ
Feeding mode of *Anopheles* larvae

4. オキナワヤブカ 幼虫の摂食方法

動画 4J

上映時間
4 分

オキナワヤブカ *Aedes o. okinawanus* の幼虫は奄美大島，沖縄島，石垣・西表島の森林内の樹洞の水溜まりに普通に見られる．成虫は胸背に褐色の線状縦斑を有し，森林内で昼間よく吸血に飛来する．幼虫の摂食の仕方は特徴的である．水底の堆積物を口刷毛でかき混ぜ，収集かじり取り法で摂食しながら微粒な堆積物を水中・水面に放射拡散させる．また，水面で懸垂静止している幼虫は頭部を内側に曲げ，口刷毛で水流を起こし，水中を浮遊する微粒子を口元に引き寄せ収集濾過法で摂食する．この蚊は分類学上ヤブカ類（*Aedes*）であるが，繊細な口刷毛を有し，イエカ類（*Culex*）とよく似た摂食方法が観察された．

Feeding mode of *Aedes o. okinawanus* larvae

Video 4E

Show time
4 m 14 s

Larvae of *Aedes o. okinawanus* are commonly found in the pools of tree holes in the forests of islands in the Ryukyu Archipelago. The adults are covered with characteristic brown linear streaks on the thoracic dorsum; they often come to feed on human blood in the forest during the daytime. The feeding mode of the larvae is characteristic, employing two feeding modes: collecting-gathering at submerged surfaces by stirring the sediment with their mouth-brushes, and supplementing this with collecting-filtering in the water column at the water surface. Although this mosquito is taxonomically classified as genus *Aedes,* the larva like *Culex*, has long, fine, unserrated mouth-brushes, comparatively large maxillae, and weakly sclerotized mandible, and it employs two feeding modes.

収集かじり取り法と収集濾過方法で水中の微細な有機物を摂食するオキナワヤブカ
Aedes o. okinawanus feeding on particles by collecting gathering and collecting-filtering mode

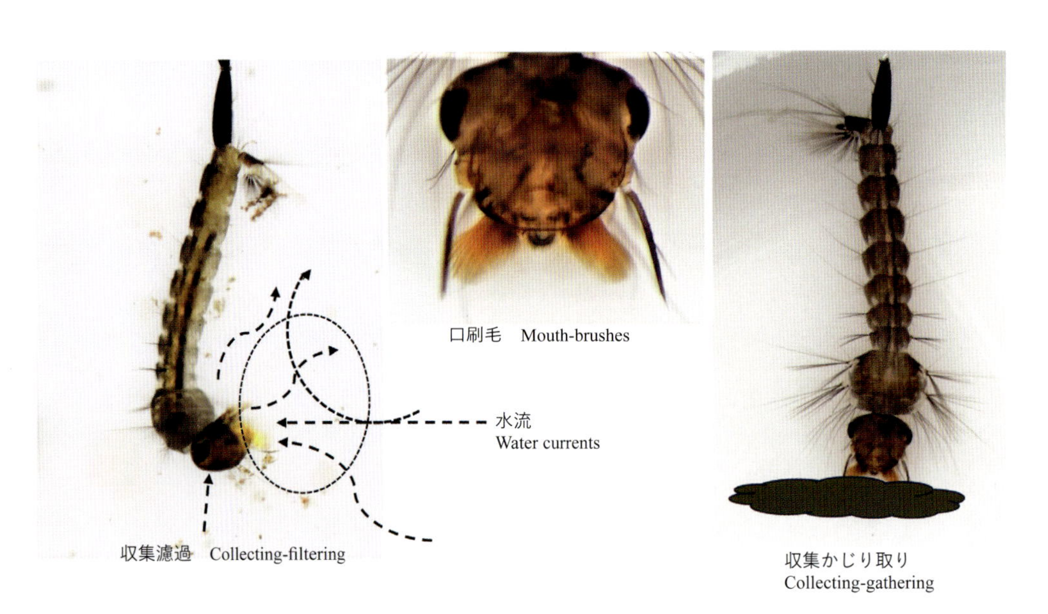

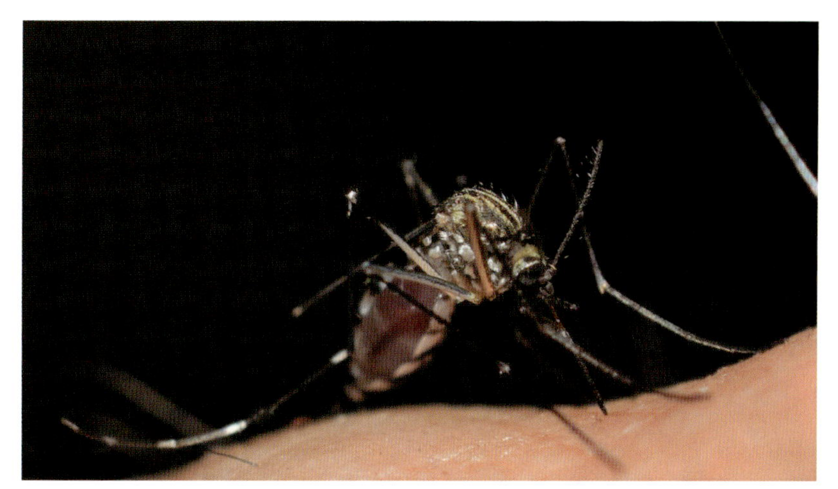

森林内で人の血を吸うオキナワヤブカ
Ae. o. okinawanus, feeds on human in the forests

森林内の樹洞の水溜まりに発生するオキナワヤブカ
Tree hole, breeding pool of the larva

5. 岩礁の塩水溜まりに多発するトウゴウヤブカ

動画 5J

上映時間
5 分 19 秒

トウゴウヤブカ *Aedes togoi* は，日本列島，東南アジアに広く分布する著名な蚊である．人のフィラリア（マレー糸状虫）の媒介蚊としてもよく知られている．幼虫は 4 月～ 7 月に海岸の岩礁の窪みや人工容器などの塩分を含んだ水溜まりに多数発生する．岩礁の窪みで採集した幼虫と溜まり水，堆積物を水槽（水深 5 cm）に移して観察した．幼虫は水面と水底の堆積物の間を頻繁に行き来し，頭部を堆積物に突っ込み口刷毛でかき混ぜ物色する様子が見られた．また，しばしば堆積物（緑藻や動物の死骸）をくわえて水面に浮上し，呼吸をしながら持ち上げた塊を口刷毛で風車のように回転させ，摂食する光景が見られた．この摂食方法は収集かじり取り法と呼ばれ，多くのヤブカ類に見られる．

Tidal brackish water pool breeder, *Aedes togoi*

Video 5E

Show time
5 m 28 s

Aedes togoi is widely distributed in the Japanese Archipelago, Korea, South China, Taiwan, Thailand and Malaysia. It is a carrier of the filarial worm *Burugia malayi.* The larvae usually occur in tidal pools or rock pools with salt or brackish water from April to July in Okinawa. The larvae were transferred with brackish water and sediment, including green algae and animal carcasses in the rock pools, to a small aquarium for observation of their feeding habits. The larvae move back and forth frequently between the water surface and the sediments, thrusting their head into the sediment and stirring them with their mouth-brush. They put small fragment of the sediment in their mouth and rise to the surface of the water, where they rotate and eat the clump. They are browsers having shorter mouth-brushes with the middle ones serrated distally, smaller maxillar and stronger mandible. The feeding mode is called “Collecting-gathering”, which is commonly found in many *Aedes* larvae.

トウゴウヤブカ雄
Adult male

胸側模様
Lateral aspect of thorax of female

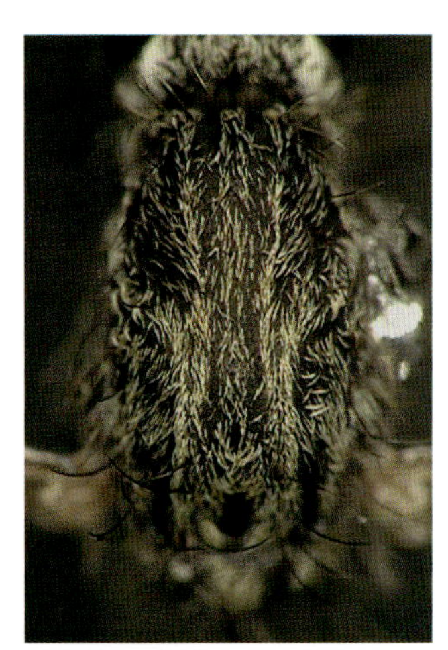

胸背模様
Thorax, dorsal aspect

トウゴウヤブカの生息地
沖縄島具志頭海岸
Breeding place, Gushikami coast
dotted with reefs

蛹　Pupa

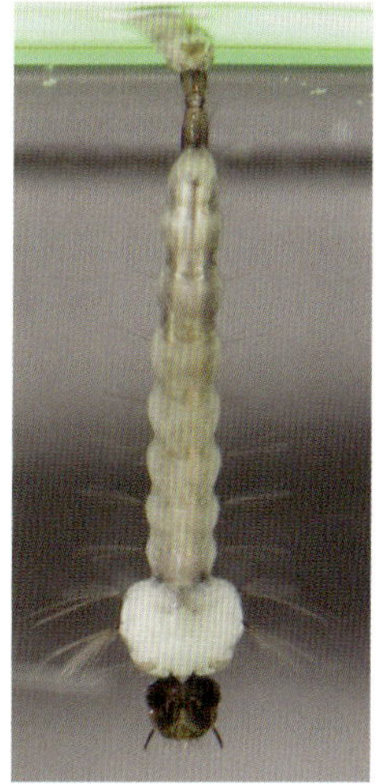

幼虫　larva

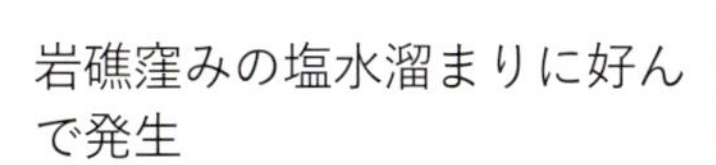

岩礁窪みの塩水溜まりに好んで発生
Saltwater pools in reef depressions

6. 緑藻・アオミドロを裁断し，摂食するカラツイエカの幼虫

動画 6J

上映時間
9 分 15 秒

カラツイエカの幼虫は陽当りの良い水田や溝でもっぱらアオミドロを摂食することはよく知られている．その摂食方法は特徴的である．アオミドロの一本の糸を触角で探り当て，大顎で切断・二分し，最初切断した下方の切口からかみ砕いて飲み込むように体内に取り込む．その後，休憩・脱糞，再度摂食する．糸を最後まで食べると，頭部を 90 から 180° 回転し，口元に留置いた断片の糸をもぐもぐし，体内に取り込む．このような留置きは 3，4 齢幼虫で健康なアオミドロが十分備わった環境では普通に見られた．幼虫期間は 7 から 10 日で，摂食―休憩―脱糞を繰り返している．限られた水槽の中ではアオミドロを食い尽くすと，食べ残したちぎれた糸を奪い合い，餌不足になり，生育期間が延び，小さな蛹，成虫になる．孵化した幼虫は，すぐにアオミドロの塊にもぐり込み糸をかみ切る悪戦苦闘の摂食が始まる．

***Culex bitaeniorhynchus* larvae, habitual shredder of *Spirogyra* algae**

Video 6E

Show time
8 m 22 s

Culex bitaeniorhynchus larvae are shredders in its feeding mode. They are habitual shredders, feed selectively on the filamentous green algae, *Spirogyra* sp. The feeding mode is unique. At first, their antennal setae touch on an algal filament, followed by clasping it with mouth-brushes, then biting it off into two fragments. One of the fragments is passed through between the antenna and the tip is clasped and kept for a while by the mouth-brushes. The larva then chops off another fragment by a combined action of the mandible, maxilla and dorsomentum, and swallows the chopped algae successively. This feeding seems to be without wastage of the algae. Call it “Efficient shredding mode” as seen in aquarium with plenty of fresh algae. Attempts to breed the larvae with different algae found in shallow open ponds, were unsuccessful. Fresh *Spirogyra* algae are of fundamental importance to this larva not only for its growth but also for survival.

緑藻に依存する幼虫　Larvae dependent on green algae

成虫の形態　Morphology of *Cx. bitaeniorhynchus*

腹部背節（雌）
Dorsal part of abdomen
(female)

緑藻が繁茂するクレソン畑
Watercress field with green algae

カラツイエカ幼虫の摂食方法　Feeding mode of *Cx. bitaeniorhynchus*

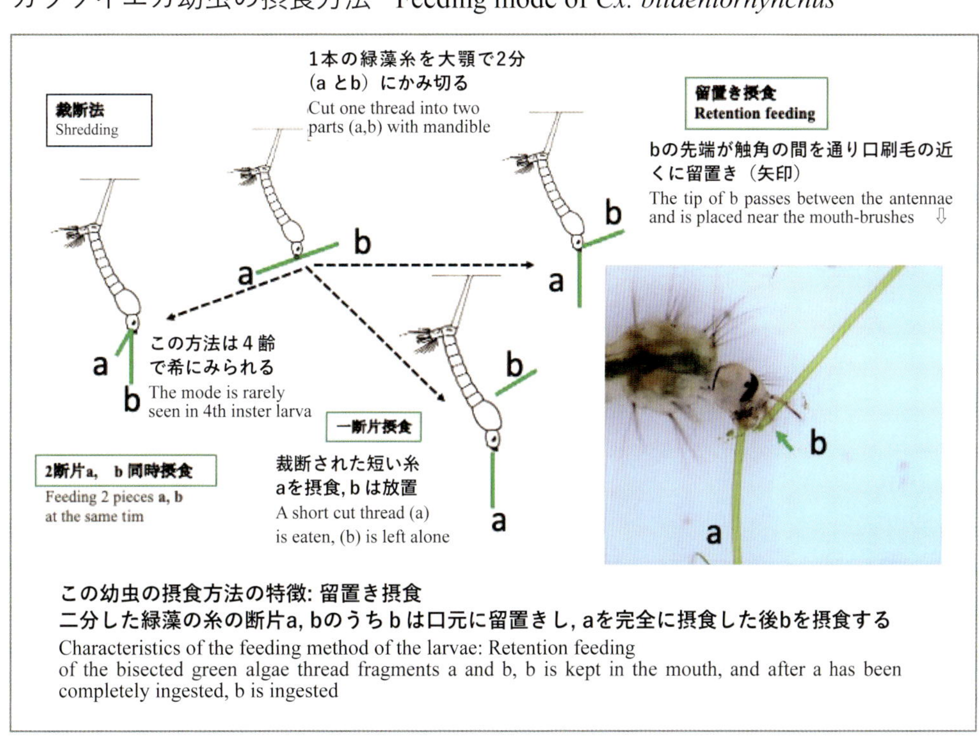

この幼虫の摂食方法の特徴: 留置き摂食
二分した緑藻の糸の断片a, bのうちｂは口元に留置きし，aを完全に摂食した後bを摂食する
Characteristics of the feeding method of the larvae: Retention feeding
of the bisected green algae thread fragments a and b, b is kept in the mouth, and after a has been completely ingested, b is ingested

7. ミナミハマダライエカ幼虫，収集濾過法と緑藻を裁断法で摂食

動画 7J

上映時間
5 分 46 秒

ミナミハマダライエカの幼虫はカラツイエカ *Culex bitaeniorhynchus* と同様，緑藻が繁茂する水溜まりに生息し，緑藻を摂食している．琉球列島での本種の発生数はカラツイエカに比較して少なく，主に清流の淀みで採集される．緑藻を摂食する方法はカラツイエカとは幾分異なり，緑藻の糸を大顎で切断後，口刷毛と大・小顎で糸をかみ砕いて体内に取り入れる．かみ切った他方の糸を口元近くに留置きする習性はみられないが，切断した 2 本の糸を束ねてかみ砕いて摂食する様子がしばしば観察された．また，緑藻を裁断摂食するだけでなく，口刷毛を盛んに震わせ水流を起こし，口元に浮遊物を引き寄せ，体内に取り込む収集濾過法による摂食も観察された．ミナミハマダライエカの口刷毛の形態は，典型的な収集濾過法で摂食するネッタイイエカと同様，長く繊細で発達している．このように本種の幼虫は，水中の浮遊物を収集濾過法と緑藻を裁断法の両刀使いで摂食している．緑藻群の中にもぐり込んでいる幼虫は，数時間水面に浮上しない．このことから幼虫は呼吸管による水表面からだけではなく，水中の溶存酸素を取り込んで呼吸をしていると思われる．

Culex mimeticus employing two feeding modes: Collecting-filtering and shredding-like modes

Video 7E

Show time
5 m 55 s

Culex mimeticus larvae, like *Cx bitaeniorhynchus* larvae, are found in freshwater stream pools overgrown with green algae, *Spirogyra* sp. In the aquarium with the algae, the larvae of *Cx. mimeticus* use two feeding modes, the “Collecting-filtering” and “Shredding” modes. The “Shredding” mode of *Cx. mimeticus* larvae is different from that of *Cx. bitaeniorhynchus;* its feeding action looks awkward. It was often observed that two algal filaments were bound together and chewed at the same time. Feeding habit of keeping an algal fragment was not observed in *Cx. mimeticus.* The mouth-brushes of *Cx. mimeticus* are long and finely developed, similar to that of *Cx. tritaeniorhynchus*, which is principally “Collector-filterers”. We believe that the feeding mode of *Cx. mimeticus* has evolved from “Collector-filterers” to “Shredder” in the environment where the algae flourish. They are not dependent on green algae as the larvae grow even in the absence of green algae.

雌成虫　Adult female

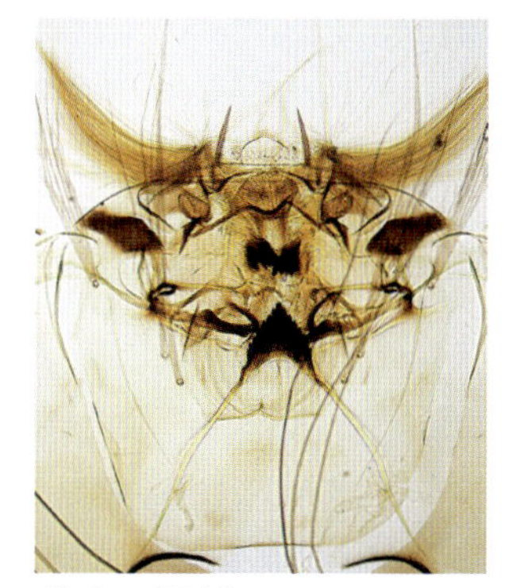

幼虫の頭部　Larval head

収集濾過法と裁断法の二刀流で摂食
Two feeding modes, Collecting-filtering and shredding were found

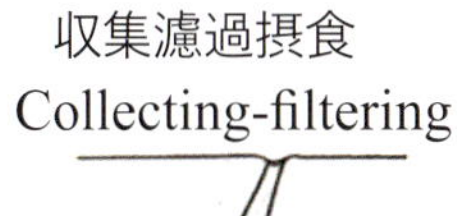

収集濾過摂食
Collecting-filtering

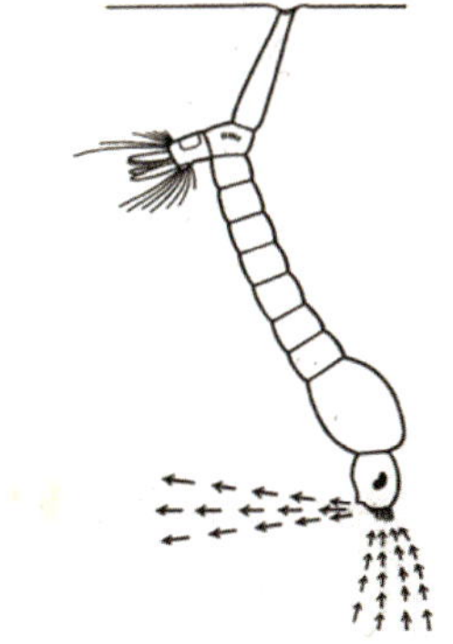

水流で浮遊物を口元に引き寄せ濾過して摂食する

The mouth-brush creates a water current, drawing floating objects to the mouth

裁断摂食
Shredding

かみ切った緑藻の糸の一方を摂食し，他方は側方に放置する

Cut algal filament into 2 parts, feed on one and leave the other

かみ切った緑藻の糸を二つ重ねて摂食する

Cut algal filament into 2 parts and feed on both pieces at the same time

幼虫の発生源　緑藻が繁茂する渓流
Larval habitat, mountain streams with green algae grow abundantly

緑藻の糸を二つ重ねて摂食
The larva feeding on two green algae threads

8. ミナミハマダライエカとカラツイエカ幼虫のアオミドロ摂食方法の比較

動画 8J

上映時間
7 分 04 秒

琉球列島ではミナミハマダライエカの幼虫はカラツイエカの幼虫と同様，アオミドロが繁茂する水溜まりに生息している．水槽での観察で，ミナミハマダライエカ幼虫の摂食は口刷毛で水流を起こし，水中の浮遊物を集めて飲み込む収集濾過法とアオミドロをかみ切る裁断法の二刀流が観察された．ミナミハマダライエカの緑藻を裁断摂食する方法はカラツイエカとは幾分異なり，頭部を振りながら緑藻の糸を大顎で無理やり切断し，口刷毛と大・小顎で糸をかみ砕いて体内に取り込む様子はカラツイエカに比べてぎこちなさを感じた．切断した 2 本の糸片を束ねて同時にかみ砕いて摂食する様子がしばしば観察された．カラツイエカにみられるかみ切った糸を口元に留置きする習性はみられなかった．ミナミハマダライエカの口刷毛の形態は，収集濾過法で摂食するコガタアカイエカやシロハシイエカと同様長く繊細で発達しており，摂食は収集濾過からアオミドロが繁茂する環境下では裁断する方法に進化してきたものと考える．ミナミハマダライエカの幼虫はカラツイエカとは異なり，アオミドロに対する依存性は弱い．

Comparison of feeding habits of *Culex mimeticus* and *Cx. bitaeniorhynchus* larvae

Video 8E

Show time
6 m 48 s

Culex mimeticus larvae, like *Cx. bitaeniorhynchus* larvae, are found in freshwater stream pools overgrown with green algae, *Spirogyra* sp. In the aquarium with the algae, the larvae of *Cx. mimeticus* use two feeding modes, the “Collecting-filtering” and “Shredding” modes. The “Shredding” mode of *Cx. mimeticus* larvae is different from that of *Cx. bitaeniorhynchus*; its feeding action looks awkward. It was often observed that two algal filaments were bound together and chewed at the same time. Feeding habit of keeping an algal fragment was not observed in *Cx. mimeticus*. The mouth-brushes of *Cx. mimeticus* are long and finely developed, similar to that of *Cx. tritaeniorhynchus*, which is principally “Collector-filterers”. We believe that the feeding mode of *Cx. mimeticus* has evolved from “Collector-filterers” to “Shredder” in the environment where the algae flourish. They are not dependent on green algae as the larvae grow even in the absence of green algae.

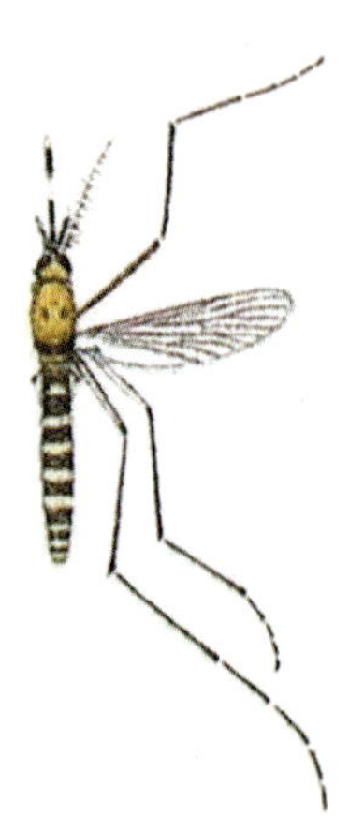

カラツイエカ雌
Culex bitaeniorhynchus female

ミナミハマダライエカとカラツイエカ
Culex mimeticus* and *Culex bitaeniorhynchus

ミナミハマダライエカ *Culex mimeticus*

翅
Wing venation

口刷毛
Mouth-brushes

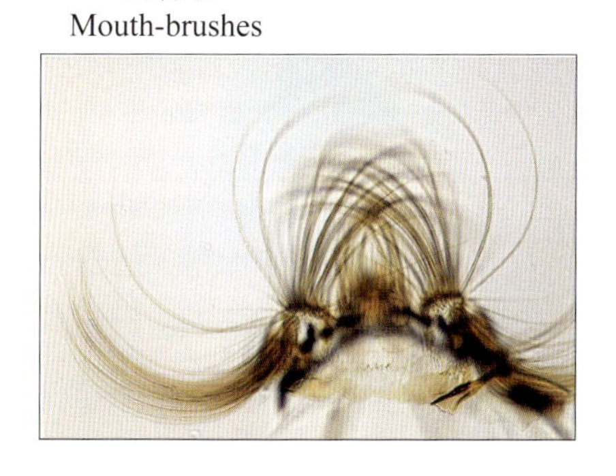

下唇基板
Dorsomentum

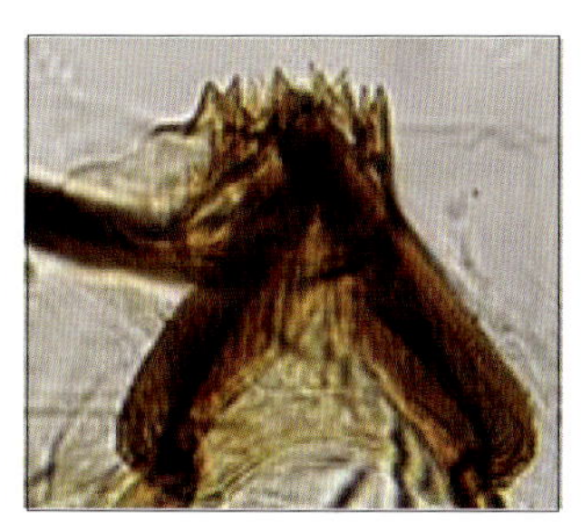

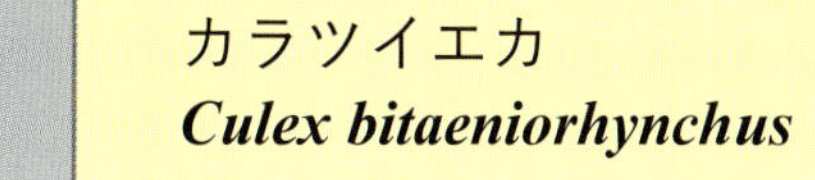

カラツイエカ *Culex bitaeniorhynchus*

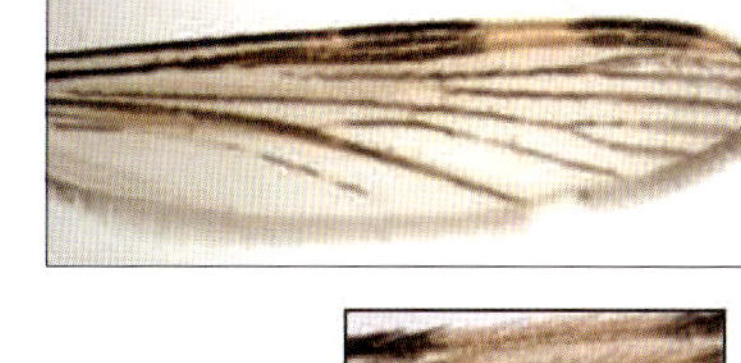

翅
Wing venation

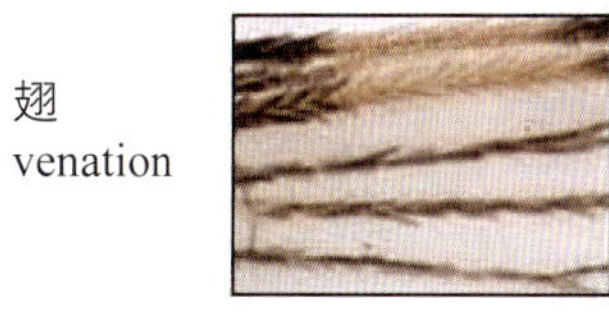

口刷毛
Mouth-brushes

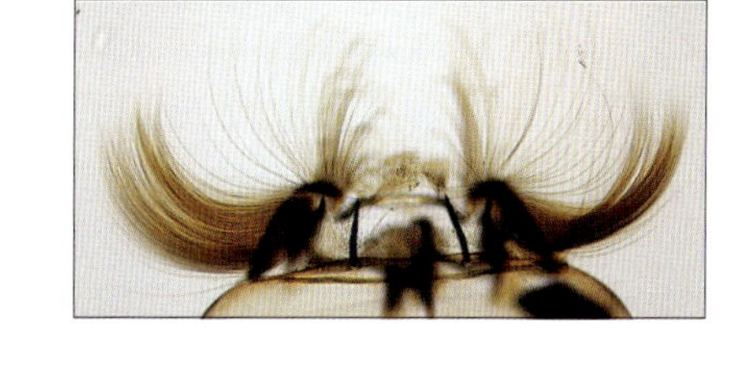

下唇基板
Dorsomentum

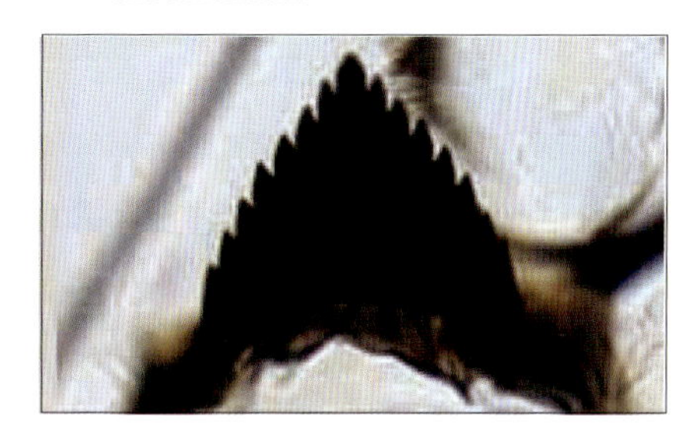

9. 湿地性植物の根に付着し，呼吸するアシマダラヌマカの幼虫

動画 9J

上映時間
12 分 34 秒

アシマダラヌマカの幼虫は湿地帯に生息し，成虫は真昼でも人の血を吸いに飛来する．吸血した蚊は 3 日後，浮き草の葉に飛び降り，触角で葉の表面を探索するように徘徊する．その後，腹部先半分を水中に沈め 100 個ほどの先が尖った卵粒を円形の卵塊にして葉の裏面に産み付ける．卵粒は約 56 時間後，先端が三角形のキャップ状に切り開かれ，幼虫が躍り出る．幼虫は適当な根を求めて遊泳し，呼吸管の先端を根に刺し入れて呼吸し，収集濾過法で摂食する．幼虫と蛹の脱皮も呼吸管・角を根に刺し入れたまま組織内の酸素を吸収し，水面には余程の振動がない限り浮上しない．蛹化の様子は劇的で，出現したばかりの呼吸角で根に突き刺さった呼吸管を挟み，抜きとり，同じ穴に呼吸角を刺し入れる．蛹化開始から幼虫の脱皮殻を脱ぎ捨てるまでの時間は個体により異なり 10 ～ 30 分だった．幼虫と蛹の呼吸法は呼吸管・角で根の組織内の酸素を吸収する通説に反して，呼吸管・角から鰓を出して水中の溶存酸素を吸収する説（岩城，1989）がある．しかし，この岩城の「鰓呼吸説」を支持する論文は知る限り見当たらない．琉球列島にはアシマダラヌマカと同様水生植物の根から酸素を吸収するムラサキヌマカ *Coquillettidia crassipe*s とキンイロヌマカ *Cq. ochracea* が生息しているが，発生数が少なく，希に人に吸血飛来する．

***Mansonia uniformis* larva attached to water plant**

Video 9E

Show time
13 m 32 s

The females of *Mansonia* glue their beautiful round shaped egg-raft to the undersides of leaves of floating plants. About 100 larvae come out directly into the water. The larvae and pupae show highly specialized respiratory behaviors. The tip of the siphon and trumpets is modified into a sharp piercing organ which is thrust into the air-spaces of submerged plant to obtain oxygen, without coming up to the water surface to breathe. They stay there for about 2 weeks and feed on suspended matter by the collecting and filtering method. The larval moulting and pupation are performed with siphon and trumpet inserted into the root in the water. During pupation, the trumpets appear first. The pupa pulls out the siphonal skin by the trumpets. After taking off the siphon, the pupa inserts its trumpets into the same puncture on the root. The emergence takes place like other species on the surface of water. The peculiar behaviors may be advantageous in avoiding being attack by natural enemies and avoiding food competition with other mosquito larvae. In the Ryukyu Archipelago, *Coquillettidia crassipe*s and *Cq. ochracea*, which absorb oxygen from the roots of aquatic plants like *Ma. uniformis*, have been recorded, but their numbers are small, and they rarely fly to humans to suck blood.

夕方，人に吸血飛来した
アシマダラヌマカ
Female fly to suck human blood in the evening

アシマダラヌマカとキンイロヌマカ
Mansonia uniformis* and *Coquilletiidia crassipes

幼虫が生息する湿地　Wetlands where the larvae live

水生植物の葉の裏面に産み付けられた卵塊
Egg raft laid on the underside of the leaves of an aquatic plant

呼吸管を根に刺し入れて呼吸する幼虫
Larvae breathe by inserting siphon into the root of aquatic plants

呼吸角を根に刺し入れて呼吸する蛹
Pupa breathes by inserting trumpet into the root of aquatic plants

キンイロヌマカの幼虫
Coquillettidia crassipes larva

キンイロヌマカの成虫
Cq. crassipes adult female

アシマダラヌマカの成虫
Ma. uniformis adult female

●アシマダラヌマカ幼虫と蛹の動きと動画撮影の工夫（水田英生）

アシマダラヌマカ幼虫の動画撮影は次のような条件の下でおこなった.

1. 観察用の水槽は葉がついた水生植物の根の向きを自由に変えられるようにやや大きめの水槽（円筒形の 500 ml 用ペットボトルを利用）を準備した.

2. 水槽には組織内に酸素を貯める活の良い若い，白色の根を選んで入れること．植物が枯れると根の組織内の酸素が少なくなり，幼虫や蛹は呼吸困難で死亡する.

3. 幼虫（4 齢）は呼吸管を根に穿刺し（写真 A)，口刷毛で水流を起こし，水中を浮遊する微細な有機物を収集濾過法で摂食するため，水槽内の飼育水は幼虫が発生している湿地の新鮮な水と時々入れ替えること．時間が経つと水が汚れて原虫などが幼虫や植物の根にべったり付着し，幼虫は衰弱する.

4. 蛹化は水中で普通夜間に行われる．最も劇的な蛹化の瞬間を撮影するため，予め蛹化開始時を推測し，シャッターチャンスに備えた．蛹化開始前日の幼虫は黒色になり，摂食を止め（口刷毛が動かない)，多少の刺激にも反応しなくなり，頭部が徐々に膨らみ，大きくなる．この時に初めて照明を点灯し，フォーカスを合わせて，撮影を開始する.

ヌマカ類の水中での蛹化は実に巧妙で，瞬間のでき事である．幼虫の呼吸管の先端にある穿孔のためのノコギリ状の組織（写真 B）は，蛹の呼吸角にはない（写真 C)．そのため蛹は呼吸角を新たに根に穿刺することが出来ないので，幼虫が呼吸管を刺し込んだ同じ孔に呼吸角を刺し替える（動画 9 参照)．この刺し替えはよく観察すると，次のような手順で行われる．蛹化（脱皮）が始まり，蛹の頭部が出現すると（写真 D)，頭部を前後左右に振りながら，反り返る（写真 E)．この動作を反復し，左右一対の呼吸角の先端が呼吸管（植物に刺し込まれている）の付け根（孔）に接触する．やがて呼吸管の両側に沿って一対の呼吸角の先端を刺し入れ，幼虫の呼吸管を挟みながら呼吸角の先端を左右に開くと蛹は固定される（写真 F)．同時に挟まれていた呼吸管はやや湾曲した形状の呼吸角に押し出され，根に食い込んで呼吸管を固定していた鋸歯が外され，幼虫の抜け殻は根から離脱し，同じ位置（孔）に呼吸角が固定される（写真 G)．脱皮は呼吸角を組織に刺し込んだまま，腹部の収縮を繰り返し幼虫の殻を脱ぎ捨てる.

ヌマカ類の幼虫や蛹は水面で呼吸するイエカ，ヤブカやハマダラカと異なり，なぜわざわざ水生植物の根に呼吸管・角を刺して組織内の酸素を吸収して生活するのだろうか？　動画を見ていただいたように，水生植物の根に呼吸管や呼吸角を穿刺する動作だけでなく，根からの酸素吸収方法は不利なことが多いと思われる．あえて有利と思われる点は，呼吸のために遊泳する必要がなく，水中で静止して根に擬態していること．このことは捕食性天敵に襲われる機会が少ないと思われる．また，ヌマカの水底での収集濾過法による摂食は水面で呼吸し，収集濾過法で摂食する他の蚊との競合がないと思われる.

今回のアシマダラヌマカの採集地は，沖

縄島中部の海岸近くの潮風が通る照間の湿地で，そこに発生する水生植物は，空心菜（エンサイ），ナンバンツユクサ，ウォータークローバー，背丈の低い単子葉植物などで，ガマやアシのような大型の植物やホテイソウは見られない．水深は 1 m 以下で底に 10 cm 程度の泥が堆積している．近年，湿地等の環境の変化が激しく，宅地化されこの地域でのヌマカの発生は年々減少している．数年後には，この地でのヌマカの採集は困難になると思われる．

白い根に呼吸管を挿入して呼吸する幼虫
Larvae breathing by inserting siphon into young root of aquatic plants

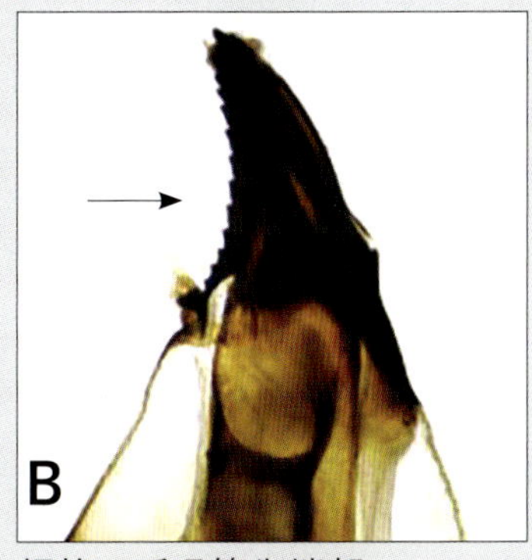

鋸状の呼吸管先端部
Sawtooth of apical part of siphon

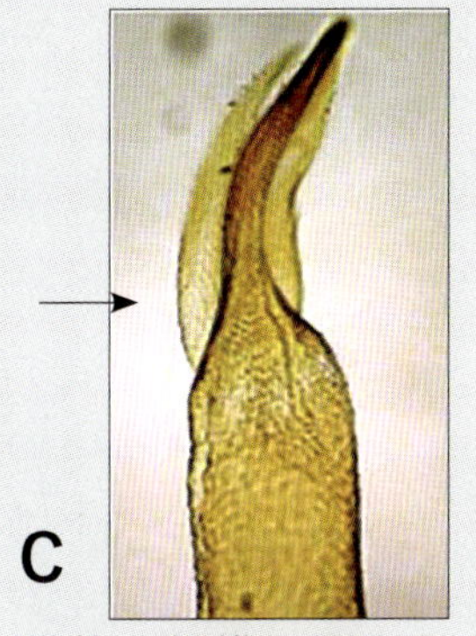

膜状の組織で覆われた呼吸角先端部
Apical part of trumpet covered with membranous tissue

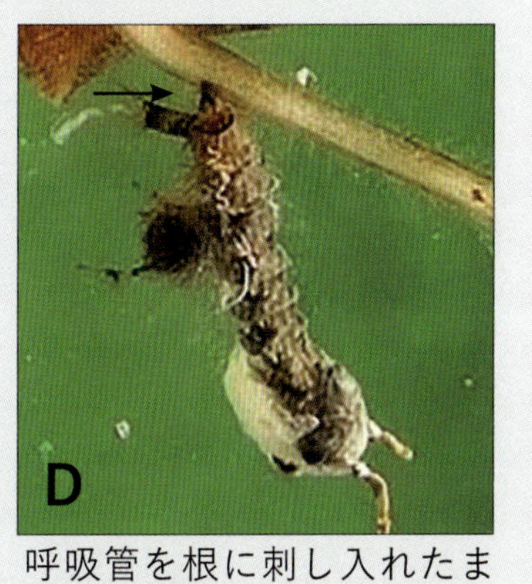

呼吸管を根に刺し入れたまま蛹化開始し，呼吸角が出現
Pupation begins with the siphon still inserted into the root, and the trumpets appear

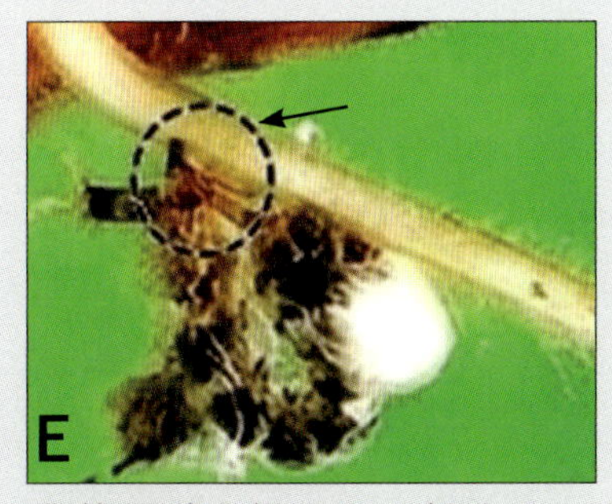

屈伸運動を繰り返し蛹化する幼虫
The larva molts its skin and pupates while repeatedly bending and stretching

呼吸角を根に刺し入れ呼吸
Pupal trumpet inserted in root for respiration

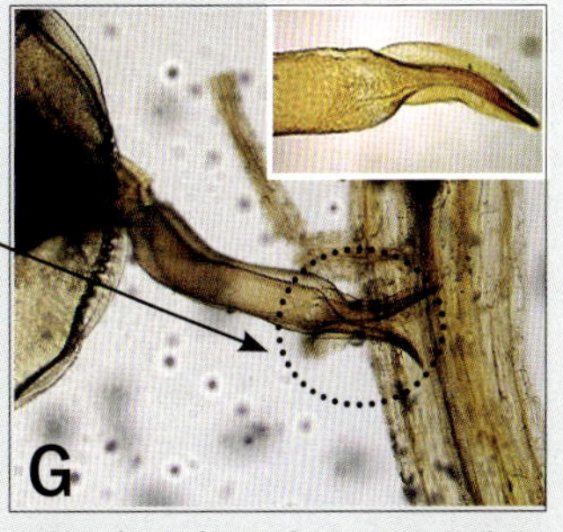

呼吸角を根に刺し入れ呼吸
Pupal trumpet inserted in root for respiration

Movements of the larvae and pupae of *Mansonia uniformis* and ideas for video recording of their behaviors

1. A slightly larger aquarium, using a cylindrical 500 ml plastic bottle, was prepared for observation so that the direction of the roots of aquatic plants with leaves could be freely changed.

2. Choose aquatic plants with few thin, and hair-like roots for the aquarium – the reason is that thin and hair-like roots get in the way when taking photos, and when the plant dies, the oxygen in the root tissue decreases, causing the larvae and pupae to die from respiratory difficulties.

3. The larvae (4th instar) puncture the roots with their breathing siphon (Photo A), create water flow with their mouth-brushes, and collect and filter fine organic matter floating in the water to feed, so the rearing water in the aquarium is occasionally replaced with fresh water from the wetland where the larvae are breeding. Over time, the water becomes dirty and protozoa cling to the larvae and plant roots, weakening the larvae.

4. Pupation usually occurs at night. To capture the pupation, estimate the time when pupation will begin and prepare for the photo opportunity. The day before pupation begins, the larva turns black and stops feeding and becomes unresponsive to even minor stimuli, and its head gradually swells and becomes larger. At this point, the light is turned on for the first time, the focus of the camera is adjusted, and the photography begins.

The underwater pupation of *Mansonia* mosquito is a very dramatic and momentary process. Unlike the larva that drills a hole by the tip of the siphon, the tip of the pupal respiratory trumpets does not have the sawtooth tool (Photo B, C) for drilling a hole in the root tissue, so the pupa cannot insert the trumpets into the root again. Therefore, the pupa inserts its trumpets into the same hole where the larva inserted its siphon (Photo, D). If we observe closely, the process of replacing the larval siphon and the pupal trumpet is as follows (Video 9E): when pupation begins and the pupal head appears, the pupa arches back, shaking its head back and forth and from side to side. By repeating this movement, the tips of the trumpets come into contact with the base (hole) of the siphon that is inserted into the plant root (Photo E). Eventually, the tips of the pair of trumpets are inserted along both sides of the siphon, and while pinching the siphon, the tips of the trumpets are opened to the left and right, fixing the pupa in place. At the same time, the siphon that had been pinched is pushed out into the trumpet, the sawtooth of the larva that had been digging into the plant root and holding the siphon in place is released, the larval exuvia is detached from the root, and the trumpets is fixed in the same position (hole) (Photo F). The larval molting is accomplished by repeated abdominal contractions, during which the pupal trumpets remain inserted into the plant tissue (Photo G).

Why do *Mansonia* larvae and pupae, unlike *Culex*, *Aedes* and *Anopheles* larvae that breathe at the water surface, go to the trouble of inserting their respiratory siphon and trumpets into the roots of aquatic plants and absorbing oxygen within the tissues to survive? As you can see from the video, not only is the act of inserting the siphon or trumpets into the roots of aquatic plants difficult, but the method of oxygen absorption from the roots is often unfavorable. What seems to be an advantage? It does not need to swim for breathing, and instead it remains stationary underwater, mimicking the plant roots. This suggests that there is less chance of being attacked by natural enemies, predators such as *Lutzia* larvae, fishes and *Cybister* water beetles. In addition, *Mansonia* larvae do not seem to compete with many other cohabiting mosquito larvae, such as *Culex* and *Anopheles*, which feed on microorganisms and detritus in the water surface by collecting-filtering, and *Aedes* larvae feeding in the bottom by collecting-gathering. In this study, *Mansonia uniformis* larvae were collected from the Teruma marshland, which is close to the coast of central Okinawajima and where sea breezes pass through. The aquatic plants that occur there are water spinach, dayflower and water clover, but no large plants such as cattails, reeds and water hyacinths can be seen. The water is less than 1 m deep, and about 10 cm of mud has accumulated on the bottom. (H. Mizuta)

10. 汚水溜まりを浄化するお掃除屋さん，オオクロヤブカの幼虫

動画 10J

上映時間
5 分 35 秒

オオクロヤブカの幼虫は，人家や森林周辺の汚水が溜まった容器に生息する．幼虫の動きは他の種とは大いに異なり，くねくねしながら，腹部先端（尾部）を先に後ずさりして遊泳（Tail-first movement）を行うほか，頭部を先に体を小刻みにくねらせ，滑るように水中や水底を素早く直進遊泳（Gliding）するなど独特な動きが見られる．汚水溜まりに堆積した有機物に頭を突っ込み，全身を震わせ 堆積物をまき散らして動きまわる．時には衰弱したボウフラや死骸を，堆積物の中からくわえて水面でむさぼり食う腐肉食者（Scavenger）である．このようにオオクロヤブカの幼虫は汚水溜まりを浄化する「お掃除屋さん」と言えるだろう．

Armigeres larva, decomposer, in polluted water pools

Video 10E

Show time
3 m 14 s

Armigeres subalbatus is an aggressive human biter. The larvae are commonly found in artificial containers with polluted water around the dwelling. It swims with characteristic translatory motion. It is achieved by a side to side lashing movement of the whole body. It moves tail-first through the water. Another common movement is gliding, head-first. The larvae propel themselves with sprinkling polluted dust by means of beatings of the long anal papilla and feeding of mouth-brushes. The larvae are not predators; they never attack and feed upon the live mosquito larvae and other insects. They are scavengers, feeding on dead invertebrates in the sediment. The *Armigeres* larvae are “Decomposers or Cleaners” in polluted water.

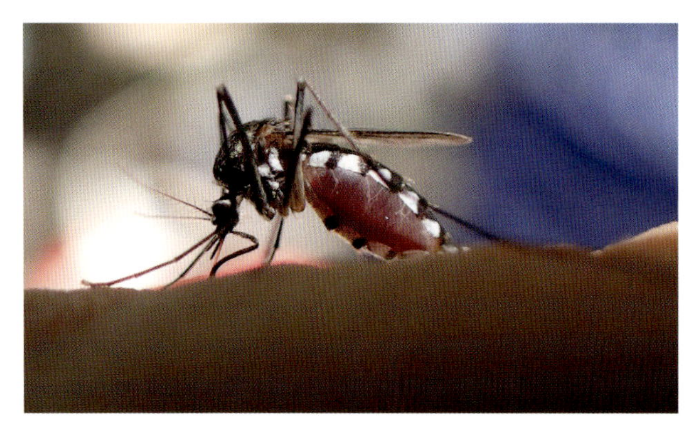

夕方人家周辺で人の血を吸う大型の蚊（雌）
A large mosquito is sucking human blood near a house in the evening

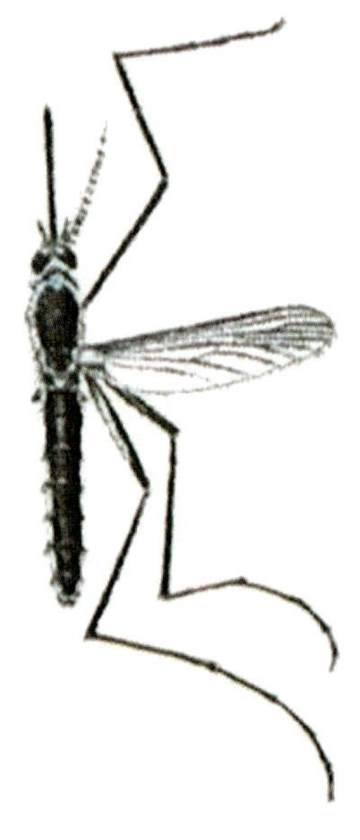

オオクロヤブカ雌
Ar. subalbatus female

汚水溜まりに好んで発生する幼虫
Larva prefers to grow in sewage pools

汚水溜まりの人工容器
Container with turbid water

汚水 Turbid water

汚水をかき混ぜ浄化する幼虫，お掃除屋さん
The cleaners, larvae stir and purify sewage

蛹　Pupae

11. ヤンバルギンモンカ の生態 I. 竹の節間の水溜まりに生息する幼虫

動画 11J

上映時間
4 分 53 s

ヤンバルギンモンカは沖縄島北部の川沿いに自生するホウライチクなどの竹の節間の水溜まりから発見された血を吸わないユニークな蚊である．この地方ではサビアヤカミキリ *Abryna coenosa* Newman の成虫が若い生（緑）竹にすり鉢状のくぼみを作り産卵する．孵化したカミキリの幼虫は竹の内側に潜入し，内壁を食べて成長する．数週間後には羽化して穴（直径 3 ～ 5 mm）から脱出する．穴が開いた竹には，竹の組織から液体がにじみ出て竹内に水が溜まり，ヤンバルギンモンカの格好な発生源となる．この水溜まりにはヤンバルギンモンカの捕食性天敵は生息せず，竹内の水は枯れることはない．温暖な琉球列島でのヤンバルギンモンカはこの格好な発生源で年中繁殖している．ヤンバルギンモンカの最北端分布地金沢での 4 年間の観察では，ホウライチクの代わりに直径 2, 3, 4, 5, 8 mm の人工的に穴をあけた孟宗竹の節間の水溜まりに幼虫が出現した．特に，3 ～ 5 mm の穴によく見られた．直径 8 mm の穴ではヤンバルギンモンカだけではなくキンパラナガハシカ幼虫も採集された．1 月の寒波の後 7 月の末まで幼虫は全く採集されなかったが，8 月～ 12 月は安定して採集された．1 月が暖冬の年は寒波が来るまで幼虫は採集されたが，寒波が厳しい年は， 翌年 8 月まで幼虫は採れなかった．積雪の極寒の金沢でもヤンバルギンモンカの幼虫は日が当たる暖かい竹やぶで，少数の幼虫が寒さに耐え，生き延び越冬し，夏季に個体数を増やしているものと思われる．

Biology of *Topomyia yanbarensis* 1. Breeding in small pools of the bamboo internodes

Video 11E

Show time
5 m 48 s

Topomyia yanbarensis is a unique non-blood-sucking mosquito that was discovered in the bamboo internodes, such as *Bambusa multiplex*, which is commonly found along the rivers in northern Okinawajima. In this region, the adult *Cerambycid* beetles make a mortar-shaped depression on young raw bamboo internodes on which to lay egg. Hatched beetle larvae bore into the bamboo and grow by feeding on the inner wall of the bamboo. After a few weeks, they emerge and escape through the hole (3 - 5 mm in diameter). Bamboo with aperture in it causes liquid to ooze out from the bamboo tissue, causing water to accumulate inside the bamboo, and making it a suitable breeding pool for *Topomyia* larvae. There are no predatory natural enemies for the larvae of this mosquito in the closed pool; moreover, the water within the green bamboo rarely dries up. The warm Ryukyu Archipelago is a prime habitat for this mosquito, where it breeds all year round. During a four-year observation in a bamboo forest in Kanazawa, the northernmost distribution area of *To. yanbarensis*, the larvae appeared in the internodes of Moso bamboo with artificial holes of 2, 3, 4, 5 and 8 mm in diameter, but mostly in internodes with 3 - 5 mm holes. In addition to *To. yanbarensis*, larvae of *Tripteroides bambusa* were also collected in the internodes 8 mm diameter hole. After the cold wave in January, no larvae were collected until the end of July, but larvae were steadily collected from August to December. Even in Kanazawa, an extremely cold and snowy area, *To. yanbarensis* larvae live in small numbers in warm, sunny bamboo forest. It is thought that the larvae survive the cold winter, and increase their population in the summer.

ヤンバルギンモンカの地理的分布　Geographical distribution of *To. yanbarensis*

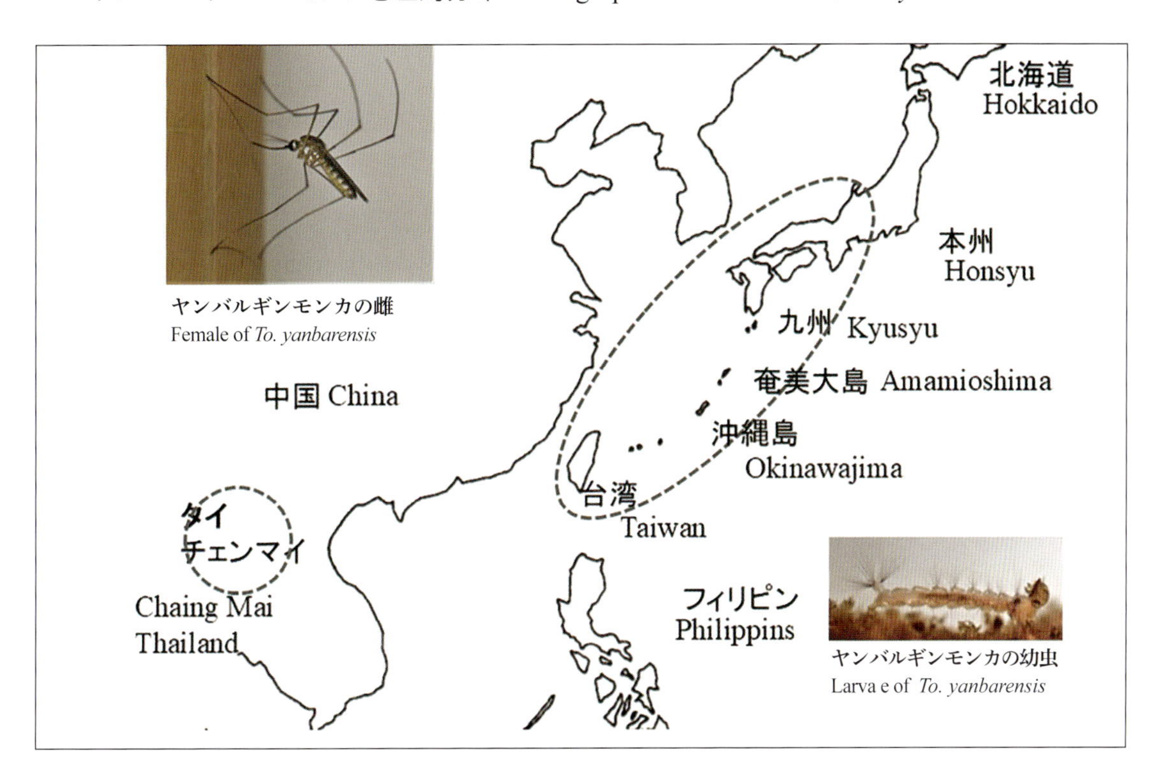

ヤンバルギンモンカの生息地ヤンバルの竹林
Habitat of *To. yanbarensis:* The bamboo forests of Yanbaru, Okinawajima

産卵のために竹に穴をあけるカミキリ
A *Cerambycid* beetle drilling holes in bamboo to lay eggs

カミキリムシに代わってドリルで竹に孔をあける
Drilling holes in bamboo instead of using longhorn beetles

12. ヤンバルギンモンカの生態 II. 交尾行動

動画 12J

上映時間
6 分 50 秒

1. ヤンバルギンモンカの交尾は竹の上で行われる.
2. 竹の上に止まっている雌成虫に，雄成虫が飛びながら斜め上の背後から近づく．雄は雌の右側から近づく場合と左側から近づく場合がある.
3. 雄は前脚と後脚を上げ，中脚は水平に開いて飛ぶ．右から接近した場合は体を落下させながら左中脚の爪を雌の右前脚の脛節に引っ掛けようとする.
4. 雄は首尾よく雌の脛節に爪が掛かると，雌の下に仰向けになり，左中脚でぶら下がる．やがて雄の羽ばたきは止み，前脚と右中脚を竹に付ける.
5. 雄の羽ばたきが始まり，雌の前脚に爪をかけて中脚を曲げ，前脚と右中脚で体を押し上げながら腹部を上方に曲げ，腹部末端の交尾器を雌の交尾器に近づける．この動作を何度も繰り返し，雄と雌の交尾器が結合すると，雄は竹から脚を放す．雌が竹につかまり，雄は反対を向き雌につながる.
6. 雄・雌はほとんど動かず，交尾成立後約 22 分で，雄が落ちるように雌から離れて交尾は終わる.

Biology of *Topomyia yanbarensis* 2. Mating habit

Video 12E

Show time
7 m 06 s

1. The mating of this species usually takes place on bamboo.
2. The male flies and approaches a female on a bamboo from diagonally above and behind. It may approach female from the right or left side.
3. The male flies with its front and hind legs raised and mid legs spread horizontally. If approached from the right, it will drop its body and try to hook the claw of its left mid leg into the tibia of the female's right fore leg.
4. When the male successfully hooks the female's tibia, it hangs supine under the female with its left mid leg. Eventually the male stops flapping its wings and its fore and right mid legs are attached to the bamboo.
5. The male begins to flap its wings, hooks its claws on the female's fore legs, bends its mid legs, pushes up its body with its fore legs and right mid leg, bends its abdomen upward, and brings the copulatory organ at the end of its abdomen closer to the female's copulatory organ. This action is repeated many times, and when the male and female copulatory organs are connected, the male releases its legs. The female clings to the bamboo, and the male faces the opposite direction.
6. The male and female hardly move, and after about 22 minutes of mating, the male falls away from the female and ending the mating.

雌雄の交尾器　Genital organs of female and male

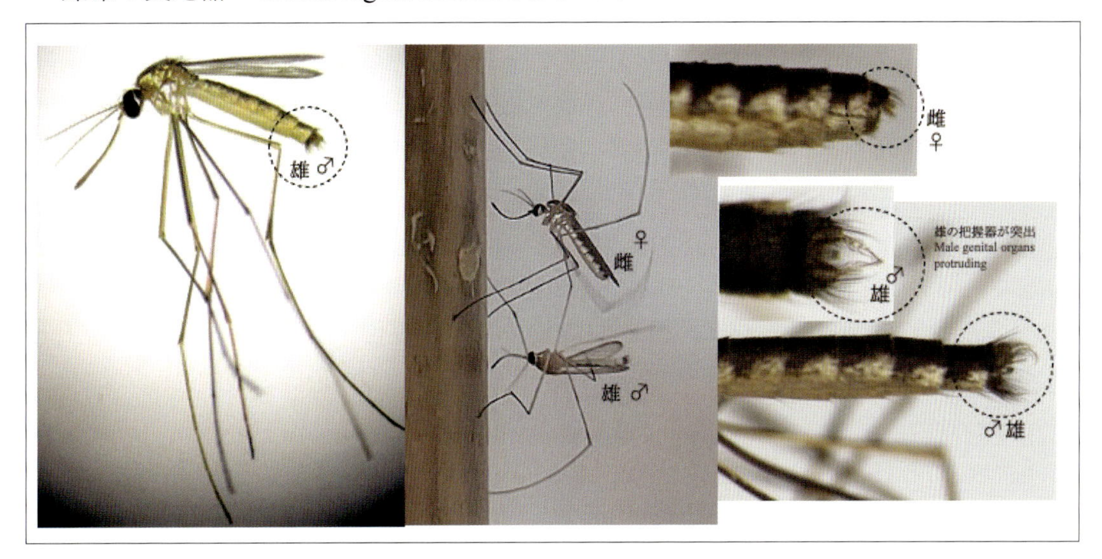

竹の上で雄の飛来を待つ雌

A female waiting on bamboo for the male to arrive

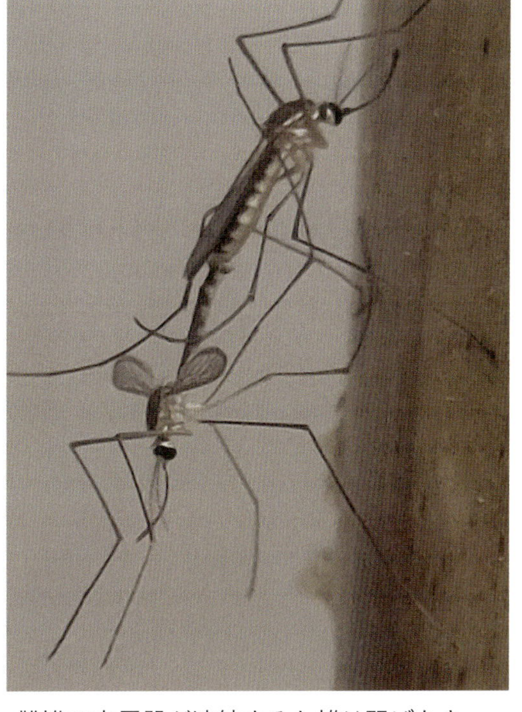

雌雄の交尾器が連結すると雄は羽ばたき，すべての脚を竹から離し、連結したまま宙づりになる

When the male and female copulatory organs connect, the male flaps his wings, releases all his legs from the bamboo, and hangs in the air while still connected

13. ヤンバルギンモンカの生態　III. 産卵行動

動画 13J

上映時間
5 分 06 秒

1. 雌成虫は吸血することなく産卵を始め，1 日に 3 ～ 5 個の卵を産む．
2. 雌は竹穴の上に止まり，前脚と中脚で体を支え後脚は常に上げている．白い卵が雌の腹部先端に出てくると羽ばたきを始める．中脚で竹をつかんだまま前脚を竹から放し上方に伸ばす．竹につかまり飛んでいる姿勢で穴に向け腹部先端を素早く振り，卵を発射すると同時に飛び立つ．
3. 卵はバラバラに孵化し，1 齢幼虫は竹の水底で成長する．
4. ヤンバルギンモンカはナガハシカ属 *Tripteroides* の「飛びながら産卵する行動」を受け継ぎ，飛ぶ姿勢で羽ばたき，中脚で竹をつかみ，体を穴の上に正確に保持することにより小さな穴にも失敗なく卵を産み込むことができるように進化した．

Biology of *Topomyia yanbarensis* 3. Oviposition behavior

Video 13E

Show time
5 m 23 s

1. *Topomyia yanbarensis* is an autogenous mosquito; the female lays eggs without sucking blood and lays 3–5 eggs each day.
2. The female perches on a bamboo hole, supporting its body with the fore and mid legs and constantly raising the hind legs. When the white egg appears at the tip of the abdomen, it begins flapping its wings. While holding the bamboo with the mid leg, it releases the fore leg from the bamboo and stretches it upward. Holding on to a bamboo in a flying posture, it quickly swings the tip of its abdomen toward the hole, at the same time launches the egg, and then flies away.
3. The eggs hatch separately and the 1st instar larvae grow on the bottom of the bamboo internode.
4. Female *Topomyia yanbarensis* shares the “behavior of laying eggs while flying” of *Tripteroides*, and by flapping its wings in a flying posture, grasping the bamboo with its mid legs, and holding its body precisely above the aperture bored by *Cerambycid* beetle, it can deposit eggs even into small aperture without failure.

穴に向かって卵を投げ入れる雌
（丸印は卵）
A female trying to throw an egg into a hole (white egg inside the circle)

竹筒内に産み落された卵，産卵直後の卵は白色
Eggs laid on inside a bamboo internode. The eggs are white immediately after laying

がじゃんコラム

【蚊帳（かや）とのかかわり】
その2．人囮二重蚊帳での蚊の採集

人や猿のマラリアを媒介する蚊の研究には人や猿を囮にした蚊帳で蚊を集める方法がある．大きな蚊帳の中に小さなベット用の大きさの蚊帳を吊りその中で囮動物は一晩過ごす．外側の大きな蚊帳の一面を開き動物に誘引された蚊を蚊帳の中に陥れる．1時間毎に蚊帳の中の蚊を手網や吸虫管で採集する．

フィリピンのパラワン島のイワヒグ囚人部落で1971年に実施した人，猿，水牛を囮にして集めた蚊を顕微鏡下で解剖した結果，猿マラリア原虫が *Anopheles balabacensis* から検出され，重要な猿マラリアの媒介蚊であることが判明した（Tsukamoto et al., 1978）．赤道直下のパラワン島で二重蚊帳の中から眺める夕方のパラワンヤリ岳（イナプ山）や満天の星空の展望は脳裏から離れない．（宮城）

フィリピン，パラワン島で動物を囮にした蚊帳によるマラリアの調査，1971

14. ヤンバルギンモンカの生態　IV. 幼虫の摂食行動

動画 14J

上映時間
4 分 42 秒

1. ヤンバルギンモンカの幼虫は，普通小さな穴があいた生竹の節間の水溜まりに棲む．
2. 幼虫の口器の口刷毛は頭の前方にあり毛の束になり，ヤブカ類によく似ているが小顎が発達している．オオカ属 *Toxorhynchites* の口刷毛の棘のような剛毛ではない．
3. 幼虫は水底に仰向けになり堆積物の中から有機物を口刷毛で集めて食べる（かじり取り，Scraping）他，水中の小昆虫の幼虫を小顎と大顎を使い捕まえてかみ砕くことなく飲み込む．オオカやカクイカ（*Lutzia*）属の積極的な捕食とは異なる．
4. 実験室ではペットとして飼育している魚の餌（® テトラミン）を粉末にしたものを与えるだけで成虫まで飼うことができる．

Biology of *Topomyia yanbarensis*　4. Feeding behavior of larva

Video 14E

Show time
4 m 58 s

1. The larvae of *Topomyia yanbarensis* are usually found in water pool in green bamboo internodes with a small hole.
2. The mouth brush of the larva is a bundle of hairs on the front of the head like *Aedes* larvae. It is quite different in the mouth brush of *Toxorhynchites* larvae which have strongly sclerotized mouth-brush and mandible.
3. The larvae can be raised into healthy adults by simply feeding them with Tetramine (gold fish powder food).
4. The larva lies on its back on the bottom of the water and feeds on organic matter in the sediment by collecting gathering or scraping mode. The larva was also seen seizing live *Culicoides* larva with its maxilla and mandible, and swallowing without chewing. The feeding method of *To. yanbarensis* preying on *Culicoides* larva is clearly different from the active predation *of Toxorhynchites* larvae.

津田良夫先生が折り紙で作成した蚊をツダギンモンカ *Topomyia tsudai* Miyagi, 2024（架空名）と名前を付けた
Miyagi named the silver mosquito that Dr. Y. Tsuda made out of origami *Topomyia tsudai* Miyagi, 2024 (Non existent name)

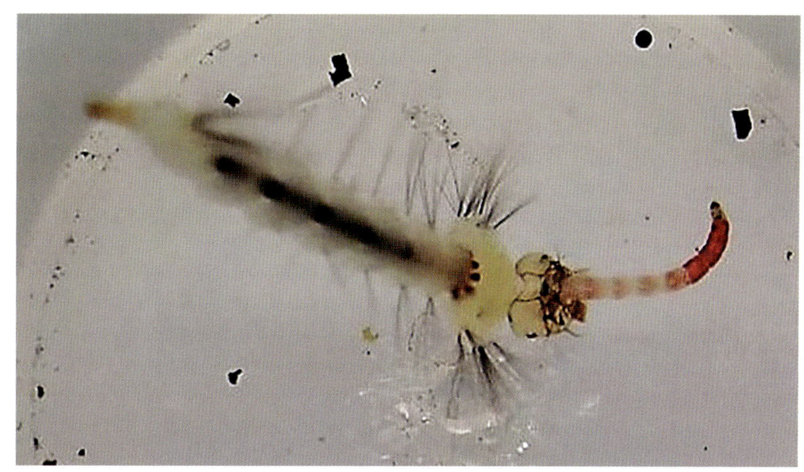

ヌカカの幼虫を丸呑みするヤンバルギンモンカの幼虫

The larvae of *To. yanbarensis* swallowing whole a biting midge larva

ギンモンカ（上段）とオオカ（下段）幼虫の口器の比較

Morphological comparison of larval mouthpart between *Topomyia* (upper) and *Toxorhynchites* (down)

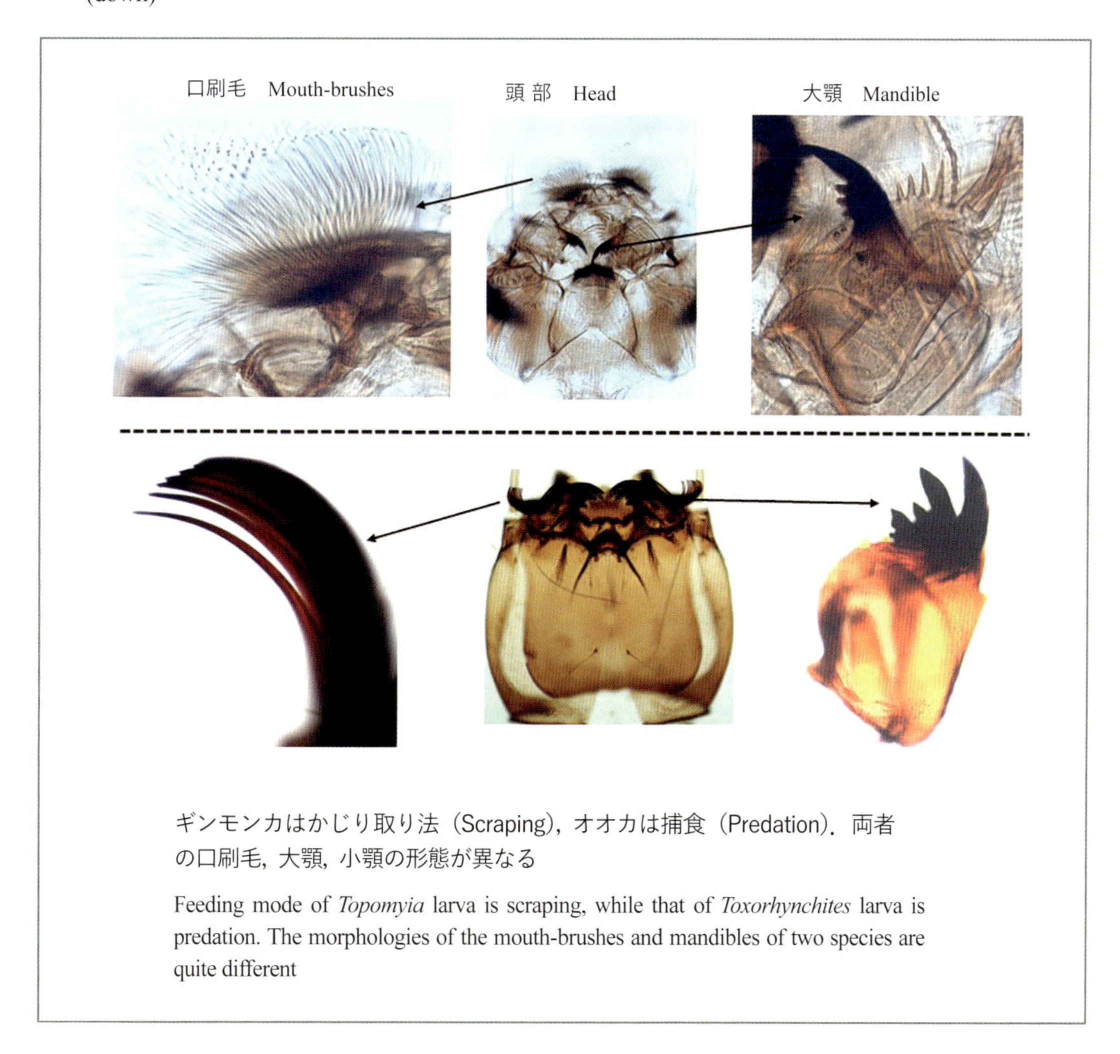

ギンモンカはかじり取り法（Scraping），オオカは捕食（Predation）．両者の口刷毛，大顎，小顎の形態が異なる

Feeding mode of *Topomyia* larva is scraping, while that of *Toxorhynchites* larva is predation. The morphologies of the mouth-brushes and mandibles of two species are quite different

●ヤンバルギンモンカの累代飼育（岡澤孝雄）

ヤンバルギンモンカは血を吸わないで産卵するいわゆる無吸血性産卵蚊である．幼虫の摂食の仕方はかじり取り法，裁断法に分類されている．水槽内では同棲する生きたユスリカやヌカカの幼虫を小顎や大顎でかみ砕いたり，丸呑みする様子もしばしば観察される．捕食性のカクイカ属（*Lutzia*）やオオカ属（*Toxorhynchites*）とは明らかに異なり，市販されている金魚の餌（テトラミン）を与えるだけで飼育できる．私は沖縄産と金沢産のヤンバルギンモンカを次のような方法で室内累代飼育を行っている．

幼虫の個別飼育

・**飼育容器**：直径 2.5 cm，高さ 4.7cm のプラスチック小容器に幼虫 1 個体を入れて個別飼育する．蛹は別の大きな容器に複数個体で飼育する．

・**幼虫の餌**：エビオス（ドライイースト），テトラミンを等分量混ぜて乳鉢で粉末にする．粉末 0.1g ＋水 9ml の混合液を 2，3 日に一度，下記の量をピペットで与える．1 齢幼虫 1 滴，2 齢幼虫 2 滴，3，4 齢幼虫 3 〜 4 滴．小容器の飼育水が濁ったら餌は控えめに与える．

・**飼育・観察箱**（25 × 15 × 35 cm）を用意する（市販のプラスチックの衣装ケースなど利用）．撮影のため全面をガラスまたは透明なアクリル板でカバーする．

観察箱には次の物（1 から 5）を設置する．

⑴ 直径 3 〜 4 cm，高さ 25 〜 38 cm の竹（下に節，上は節無し）2 〜 3 本．これらの竹の中ほどに直径 5 mm の穴をドリルであける．竹の中に水 40 〜 50 ml を入れ，竹の上部の切口は黒色のビニールシートで蓋をする．

⑵ 砂糖水（3%）を含んだ脱脂綿

⑶ 花瓶に入れた草（箱内の湿度調節）

⑷ 蛹を入れる容器

⑸ 出入口（布の吹き流し）

幼虫の個別飼育　Individual rearing of the larvae

ヤンバルギンモンカ幼虫
To. yanbarensis

Laboratory colonization of *Topomyia yanbarensis*

Topomyia yanbarensis is a so-called autogenous mosquito that lays eggs without sucking blood. The feeding methods of the larvae may be classified into scraping, browsing and shredding. They are often observed chewing up live *Ceratopogonid* larvae with their mandibles and maxillae, or even swallowing them whole. This method of feeding is clearly different from that of obligatory predators such as *Lutzia* and *Toxorhynchites* larvae, which feed on live larvae.

Topomyia larvae can be bred by simply feeding them commercially available goldfish powder food (Tetramine). Okazawa has been rearing successive generations of Okinawa and Kanazawa strains indoors using the following method.

Rearing of individual larvae: A single larva is placed in a small plastic container 2.5 cm in diameter and 4.7 cm in height for individual rearing.

Pupae: Reared together in a separate larger container.

Larval food: Mix equal amounts of EBIOS (Dried yeast) and Tetramine in a mortar and grind into powder. Feed the mixture of 0.1 g of powder and 9 ml of water once every 2–3 days with a pipette in the following amounts: 1 drop for 1st instar larvae, 2 drops for 2nd instar larvae, 3–4 drops for 3rd and 4th instar larvae. Feed sparingly if the water in the small container becomes cloudy.

Breeding and observation box (25×15×35 cm): Use a commercially available plastic clothing box, etc. Cover the front with glass or a transparent acrylic panel for photography. Place the following items (1 to 5) in the observation box. (1) Two or three bamboos, 3–4 cm in diameter and 25–38 cm tall (with nodes at the bottom and no nodes at the top). Drill a 5 mm hole in the middle of each bamboo. Pour 40–50 ml of water into the bamboo and cover the cut top of the bamboo with a black vinyl sheet. (2) Absorbent cotton soaked in sugar water (3%). (3) Grass in a vase (for regulating humidity inside the box). (4) A container for rearing the pupae. (5) Entry/exit point (Fabric streamer). (T. Okazawa)

観察箱　Observation box

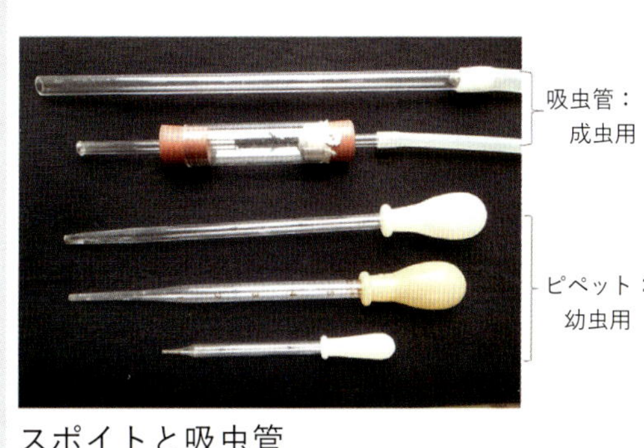

スポイトと吸虫管
Pipettes and sucking tubes

15. トラフカクイカ 幼虫の捕食 －餌食の奪い合い－

動画 15J

上映時間
9 分 58 秒

カクイカ属（*Lutzia*）の幼虫は同棲する蚊の幼虫を捕食する蚊の天敵としてよく知られている．水槽に 2 個体の空腹状態のトラフカクイカと餌食としてネッタイイエカ 4 齢幼虫数個体を入れてカクイカの捕食行動を観察した．1 個体の餌食に 2 個体のカクイカが胸部と腹部にほぼ同時にかみつき，餌食の肉汁を奪い合ってがつがつ食べた．餌食を食べつくす直前，両カクイカの口器が接触すると，後からかみついたカクイカが餌食を放棄した．先にかみついたカクイカが餌食を独り占めにし，餌食の硬い頭部と呼吸管を除いて約 2 時間 30 分で幼虫の肉汁を食べつくした．カクイカの 4 齢幼虫は蛹になるまでの 3 日間で 35 ～ 40 個体のネッタイイエカ 4 齢幼虫を捕食した．トラフカクイカ（*Lutzia vorax*）は人家周辺で有害蚊であるネッタイイエカと人工容器にしばしば同棲し，捕食する有益な天敵である．

Predators, *Lutzia vorax* larvae: Battle for prey, *Cx. quinquefasciatus* larvae

Video 15E

Show time
10 m 03 s

Two hungry predatory mosquito larvae, *Lutzia vorax* and several *Culex quinquefasciatus* larvae as their prey were introduced into a small aquarium for observation of their predatory habit. One of the *Lutzia* larvae, waiting for the prey to come near, seized it successfully. Another hungry *Lutzia* larva seized the apical part of the abdomen of the same prey at almost the same time. The two predators seized the same prey and ate it greedily from opposite ends. About 120 minutes after biting the prey, most of the prey body were consumed. The battle then began over the single prey. The two predators gradually came close together and their mouth parts made contact. The *Lutzia*, larva that bit the prey later, released it suddenly. The battle was over without any damage for the predators. The victorious *Lutzia* larva monopolized the prey and ate up its whole body except the chitinous head capsule and siphon. The *Lutzia* larvae have serrated mouth-brushes, strong sclerotized mandibles, and comparatively reduced maxillae.

2 個体のカクイカが 1 個体の餌食を奪い合って捕食
Two *Lutzia* larvae are competing a single prey mosquito

トラフカクイカ雌（♀）
Adult female *Lutzia vorax*

両種成虫の形態の違い　Morphological differences of the two species

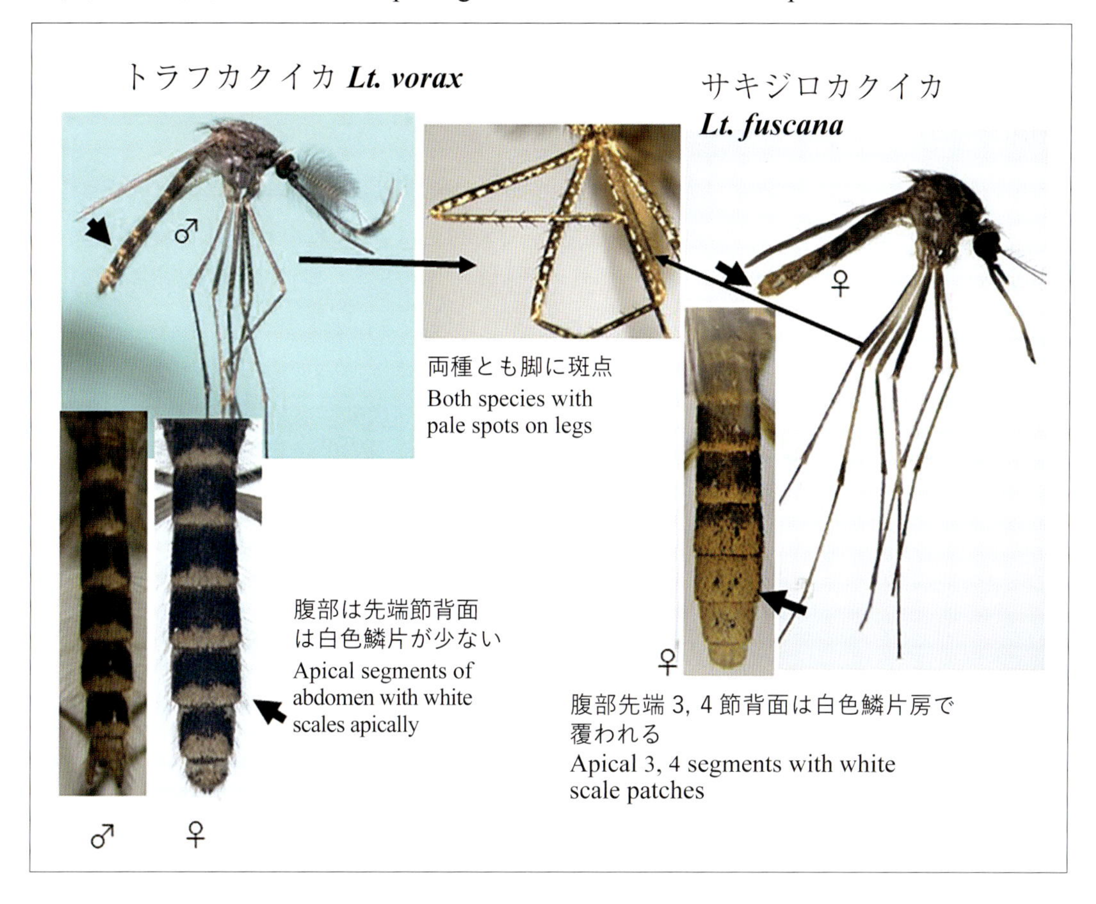

トラフカクイカ幼虫の口器　Morphology of larval mouth part of *Lt. vorax*

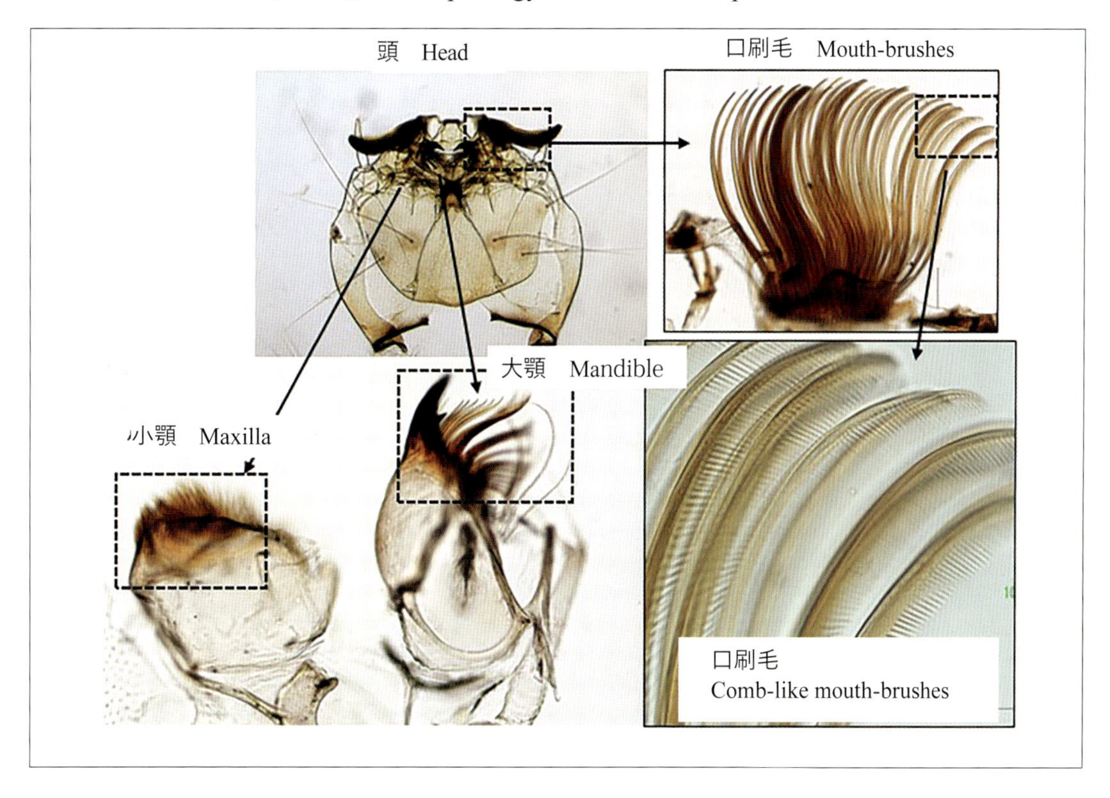

16. 樹洞の水溜まりでユスリカの幼虫を捕食するヤエヤマオオカ

動画 16J

上映時間
7 分 18 秒

水槽内での観察では，しばしば水底の堆積物上をはい回り，時には泥の中に頭部を突っ込み，物色する行動がみられる．ヤエヤマオオカは堆積物からなる筒状の巣の中に潜んでいるユスリカを見つける術を持っているようだ．

琉球列島に生息するヤエヤマオオカとオキナワオオカは比較的おとなしい性格で，樹洞の水溜まりに同棲するいろいろな蚊の幼虫だけではなく，ユスリカ，線虫の仲間を 1 日に数個体無駄なく捕食する．蛹化前に餌食を生殺し，放置するキリング（Killing）習性は見られない．

Toxorhynchites m. yaeyamae preying on *Chironomid* larvae

Video 16E

Show time
7 m 25 s

Under the aquarium condition, the *Toxorhynchites* larva is often crawling around on the mud and inserting its head into the mud looking for some foods. The larva has an ability to find out its prey lurking in the mud nest. The *Toxorhynchites m. yaeyamensis* and *Tx. okinawensis* larvae are relatively gentle and preys upon not only mosquito larvae but also other aquatic insects, such as *Chironomid* and *Ceratopogonid* larvae, and tiny nematodes on the mud. The larva does not have a killing behavior, that is to kill prey indiscriminately just before pupation.

吸蜜中のオオカ雌
Female *Toxorhynchites* sucking nectar

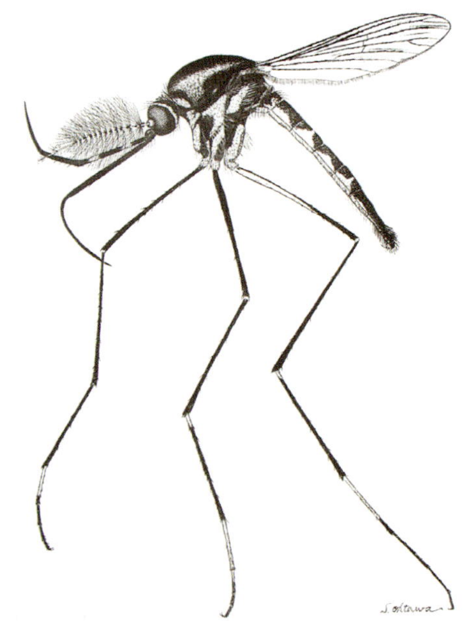

オキナワオオカの雄
Male *Toxorhynchites okinawensis*

雄の口吻　Proboscis 雄

幼虫の発生場所，樹洞
Tree hole, habitat of the larva

泥の巣の中のユスリカにかみついたヤエヤマオオカの幼虫
The larva of *Tx. m. yaeyama*e biting a midge in a mud nest

ヒトスジシマカの幼虫を捕食するオキナワオオカの幼虫
The larva of *Tx. okinawensis* preys on the larvae of *Ae. albopictus*

17. 沖縄島北部の森林内で人の血を好んで吸うオオハマハマダラカ

動画 17J

上映時間
4 分 04 秒

オオハマハマダラカ *Anopheles saperoi* は現在，沖縄島北部と西表島にだけ生息する特産種である．成虫は沖縄島北部の森林内に昼間立ち入ると最初に吸血に飛来する．昼間吸血性のハマダラカは世界的にも珍しい．森林内ではイノシシの血を吸って渓流に産卵し，繁殖する．イノシシ（吸血源動物）と清水渓流（発生源）を兼ね備えた自然環境が開発とともに消失し，この蚊の生息分布域は減少している．現在，この蚊は最初に発見された石垣島（Type-locality）では消滅したと思われる．沖縄島では年中吸血・産卵活動が見られ，降雨量が少ない冬季では渓流の水量・流が安定し幼虫が多数発生する．冬季でも気温が20℃を超えると森林内では成虫が吸血に飛来する．

Anopheles saperoi prefers to feed on human blood in the forest of the northern part of Okinawajima

Video 17E

Show time
4 m 09 s

Anopheles saperoi is an endemic species of the Ryukyu Archipelago, currently found only in the northern part of Okinawajima and Iriomotejima. When we enter the forest during the day, the mosquitoes will come to suck our blood. This mosquito of diurnal feeding habit is rare in the world. It feeds on the blood of wild boars and lays eggs in clean mountain streams. In Ishigakijima, the type-locality of the species, it is believed that this mosquito has disappeared with the depletion of wild boars and the pollution of streams due to urbanization. Blood-sucking and egg-laying activities of the females are observed throughout the year. In December to February when there is little rainfall, the water volume and flow of mountain streams are stable, and many larvae are present; when the temperature exceeds 20℃, the female adults fly to suck blood in the forest.

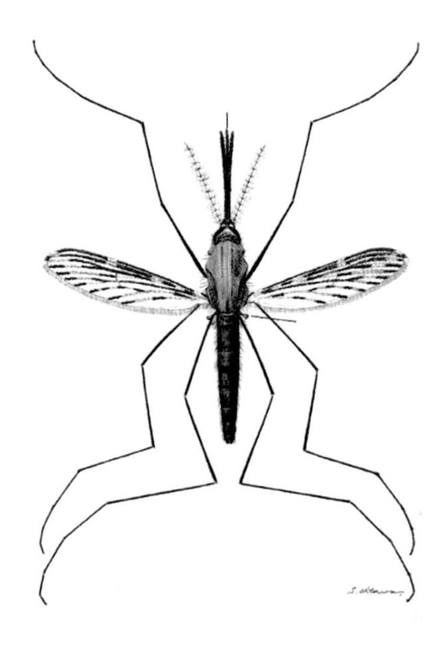

オオハマハマダラカの雌成虫
Adult female of *An. saperoi*

オオハマハマダラカの生態　*Anopheles saperoi*

腹部先端を上に斜めに静止する
Adults rest with the tip of their abdomen raised diagonally upward

幼虫の生息場所は森林内の清流の淀み
Breeding place of the larva is stream in the forest

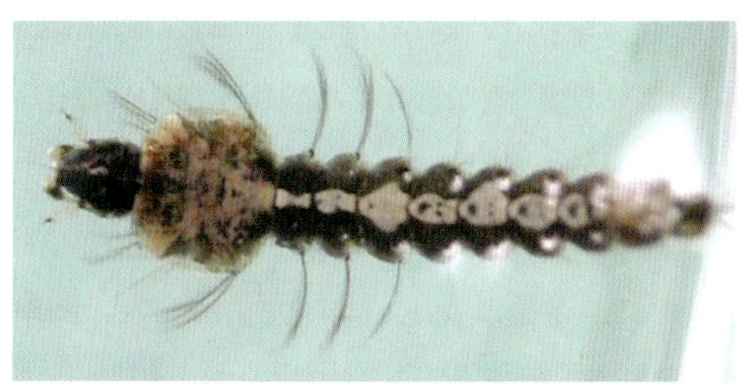

水面に横たわる幼虫，正面真上か撮影
A larva lying on the water surface, photographed from directly above

水面に平行に静止する幼虫，真横から撮影
A larva resting parallel on the water surface, photographed from directly lateral side

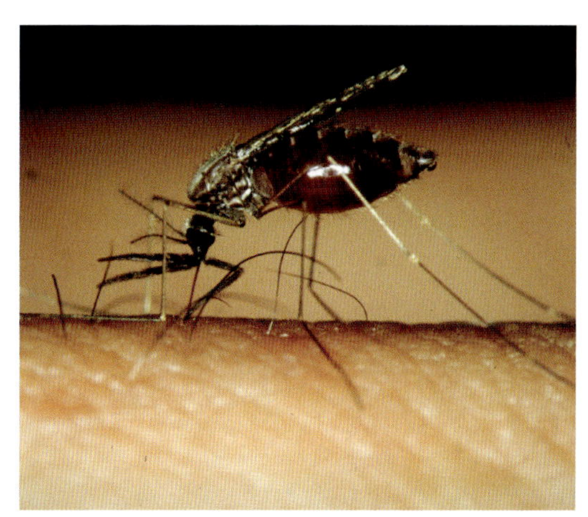

森林内で人の血を吸う成虫
Adult female that suck human blood in forests

奥土晴夫氏　撮影，西表島で

この蚊の主要な吸血源と思われているイノシシ
Wild boars are thought to be the main source of blood meal for this mosquito

18. 西表島に生息する特産種クロフトオヤブカの特異な交尾行動

動画 18J

上映時間
4 分 42 秒

1978 年 5 月，西表島古見の森林内には数日間降り続いた雨であちこちに小さな一時的な水溜まりが見られた．水溜まりの表面を注意して見ると水中には幼虫が，水の表面には蚊の成虫が群がっていることに気が付いた．交尾行動は観察箱の中で撮影された．数個体のクロフトオヤブカ *Ve. iriomotensis* の雄が水面すれすれに旋回し，羽化中の雌の体に止まり，旋回を繰り返している．次々多数の雄が羽化中の一個体の雌に触れ，雌の周りで逆立ちになり，先を争って生殖器を雌の腹部に接触し，連結しようとしている．雌が完全に脱皮すると逆立ちした一個体の雄が先を争って雌の交尾器を把握し，連結した．連結した雌雄は水面を約 30 分間漂い，やがて表皮が黒くなり，成熟した体格が良い雌は雄を引きずるようにして水面から岸辺にはい上がった．この間，別の雄がカップルに飛び掛かり交尾を仕掛けるが，連結は硬く離れることはなかった．岸にたどり着いたカップルの連結は雌の後脚のひと蹴りで解かれ，一連の交尾行動は終わった．この蚊の交尾行動開始にはある種の性的誘引フェロモンが関与していると思われる．

Peculiar mating behavior of *Verrallina iriomotensis*, an endemic mosquito of Iriomotejima

Video 18E

Show time
4 m 47 s

In May 1978, after several days of continuous rain, many small temporary water pools were seen in Komi's forest. I noticed that adult mosquitoes were swarming around the pools. A male *Verrallina iriomotensis* was circling just above the surface of the water. The male perched on the body of the female and repeated circling. One after another, many males touched a single young female during emergence, perching upside down around the female; they raced to touch and connect their genitals with the female's genitalia. When the female had completely molted, one of the males, perching upright, scrambled to grab hold of the female's genitalia to establish mating. The male and female pair in copula floated on the surface of the water for about 30 minutes, and the mature female dragged the male from the surface of the water to the shore. During this time, another male pounced on the couple and tried to mate, but the coupling remained firm. The coupling was broken with a single kick of the female's hind legs. The initiation of the mating behavior in this mosquito is thought to involve some kind of sex-attracting pheromone.

発生源は森林内に降雨後一時的に出現する水溜まりで，その水面で交尾行動が見られる
Mating behavior can be seen on the surface of puddles that appear temporarily after rainfall in forests

森林内の小道に梅雨時に出現する人や動物の足跡の水溜まりで幼虫を採集する
The larvae are collected in temporary puddles that appear on forest paths during the rainy season

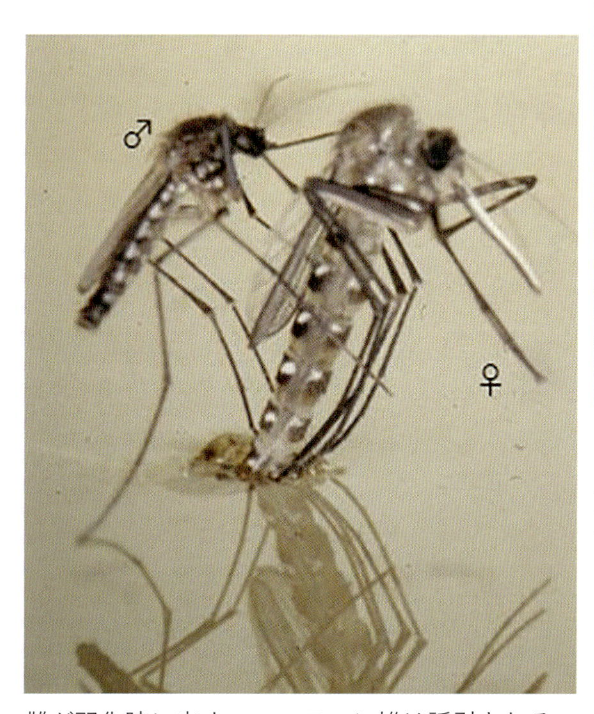

雌が羽化時に出すフェロモンに雄は誘引される
Males are attracted to the pheromones released by females when they emerge

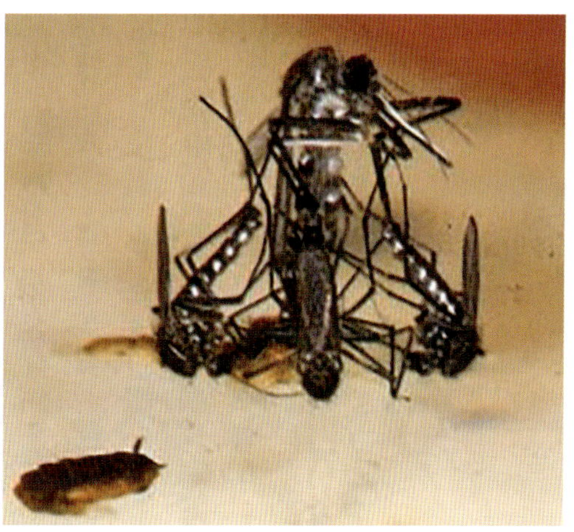

羽化中のみずみずしい雌に多くの雄が我先にと絡みつく
Many males rush to cling to the freshly emerged female

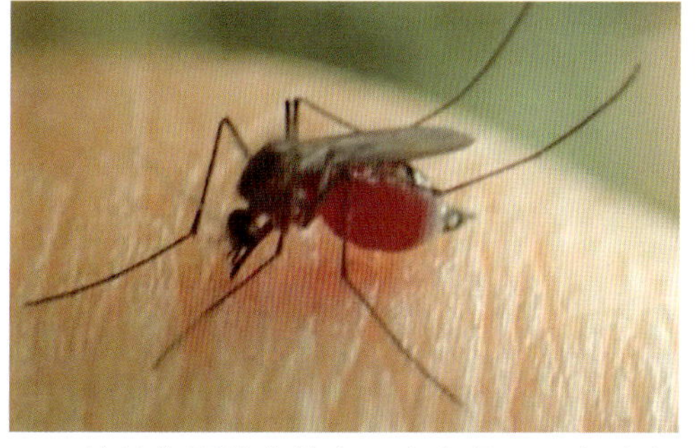

この蚊は梅雨時森林内で血を吸いにうるさくつきまとう
This mosquito persistently follows us in the forest during the rainy season for feeding blood

19. マングローブ林内で魚の血を吸うカニアナヤブカ の生態

動画 19J

上映時間
6 分 51 秒

琉球列島，西表島の河口，汽水域に広がるマングローブ林に昼間一歩踏み入れると目前に異様な光景が広がる．タコの足のような根（支柱根）に支えられ繁茂するヤエヤマヒルギ，その根元に高さ 1 メートルにもなる小火山のような泥の塚が散在している．泥の上を飛びはねながら逃げ回るデメキンのようなトビハゼが目に入る．塚の周りには小さな穴があり，大小のカニがゆっくり横歩きして穴に出入りしている．大きな塚穴の入り口に丸い小石を投げ入れると，先端の水溜まりに到達し，ポトンとはね返り音が聞こえた．坑道の先端に十分水が溜まっているあかしだ．穴の中から数個体の蚊が飛び出るが，しばらくすると必ず同じ巣穴に戻ってくる．早速，吸引機を取り出し，地下水を吸い上げると塩水に多数のボウフラが混入している．この蚊はマングローブ林内のアナジャコや周辺に生息するオカガニの巣穴だけに発生するカニアナヤブカ *Aedes baisasi* と同定された．

日暮れ再びマングローブ林の塚の前に腰を下ろし，塚の下から穴の入り口をながめると数個体のカニアナヤブカの雄成虫がスワーム（乱舞）し，交尾している様子がシルエットを描いて見える．トビハゼは縄張り争いで疲れたようで，泥の上や流木の上で休息している．時々泥の中で体を濡らし再度同じ所に戻り休息している．静寂暗黒の林内では，時折リュウキュウコノハズクの鳴き声がコホッ，コホッとあちこちから聞こえてくる．真夜中，塚の巣穴をサーチライトで照らすと，泥の塊を地下から運び巣穴の入り口に積み上げているオキナワアナジャコの赤色の頭部が一瞬現れる．泥の運搬は数分に一度，朝方まで続いた．前日石ころを投げ入れた塚の穴入り口には腹部が血液で充満したカニアナヤブカの成虫が休息している．この蚊の腹部内の血液は DNA を解析し，トビハゼ類由来の血液であることが判明している．カニアナヤブカが湿地帯でトビハゼを吸血している光景にはまだ遭遇していないが，この蚊はマングローブの汽水地帯でトビハゼ類の血を吸って，オキナワアナジャコの巣穴に生息している珍しい南方系の蚊である．このように西表島産のカニアナヤブカは室内でミナミトビハゼを吸血したことから野外でもマングローブ林内に生息する魚類，主としてトビハゼの血を吸っていると思われていた．その後，Miyake et al.（2019）は琉球列島（奄美大島，沖縄島，石垣島，西表島）から 230 個体の吸血した雌のカニアナヤブカを採集し，雌の腹部からまだ消化されていない吸血源由来の DNA の配列を決定し 4 目 8 科の魚を同定した．その結果，西表島ではジャノメハゼ（ハゼ目，ノコギリハゼ科），沖縄島と奄美大島ではゴマホタテウミヘビ（ウナギ目，ウミヘビ科）が主要な吸血源で，トビハゼ類は 7 個体（3％）であった．これらの魚はマングローブの湿地帯内をはい回る“空気呼吸魚”として知られている．

Aedes baisasi feeding on the mudskipper, *Periophthalmus argentilineatus* in the mangrove forest

Video 19E

Show time
5m 37s

One of the characteristic landscapes of mangrove swamps in the Ryukyu Archipelago, Japan, is the tall volcano-like mounds. These mounds are made by the mud lobster, *Thalassina anomala*. The animal excavates extensive burrows and lives in the underground water. The burrow system in the ground is extremely long and deep. It may have several openings and is important for the mosquito *Aedes baisasi*. The mosquito utilizes the holes of mounds as breeding sites as well as adult resting places. The females of this mosquito may feed uniquely on the mudskipper, *Periophthalmus argentilineatus* in the mangrove swamp as it feed on mudskiper in the cage. Since the female *Aedes baisasi* from Iriomotejima sucked the blood of the mudskipper in the cages, it was thought that the females also sucked the blood of fish living in mangrove forests, mainly mudskippers, even the field.

Subsequently, Miyake et al. (2019) collected 230 blood-fed female *Aedes baisasi* from different islands of the Ryukyu Archipelage such as Amamioshima, Okinawajima, Ishigakijima, and Iriomotejima, and examined DNA from the undigested blood meal source from the female abdomen. The sequence was determined and fish from 4 orders and 8 families with 15 species were identified. As the result, the main blood-sucking source for Iriomotejima was four-eyed sleeper, *Bostrychus sinensis* (Butidae, Gobiiformes), and for Okinawajima and Amamioshima, snake eel, *Pisodonopis boro* (Ophichthidae, Anguilliformes) was the main source of blood-engorged. There were 7 individuals (3%) of mudskippers. These fish are known as "Air-breathing fish" that crawl within mangrove wetlands.

マングローブ林内の塚はオキナワアナジャコの巣
The mounds in the mangrove forest are the nests of Okinawan burrowing crabs, mud lobster

オキナワアナジャコ・Mud lobster (*Thalassina anomala* Herbst)　巣穴の構造
Structure of mud lobster burrow

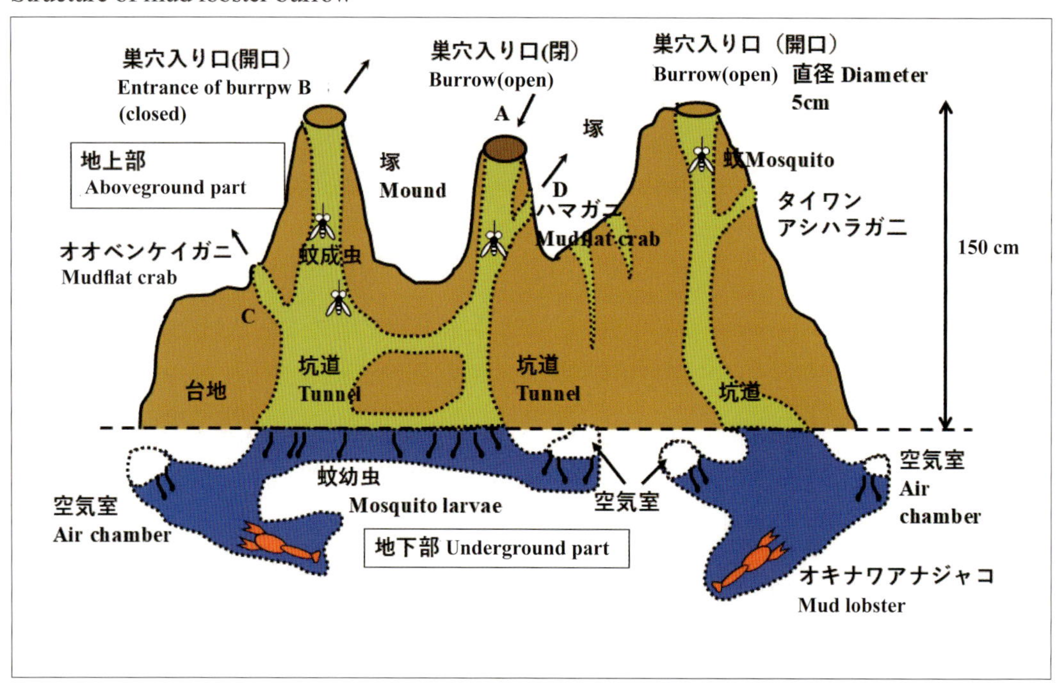

トントンミー（ミナミトビハゼ）の背中で吸血中のカニアナヤブカ
Aedes baisasi feeding on mudskipper

西表島のマングローブ湿地帯の倒木に上で休むミナミトビハゼ
A mudskipper resting on a fallen tree in the mangrove swamps of Iriomotejima

奥土晴男氏　撮影 Photo by Mr. H. Okudo

がじゃんコラム

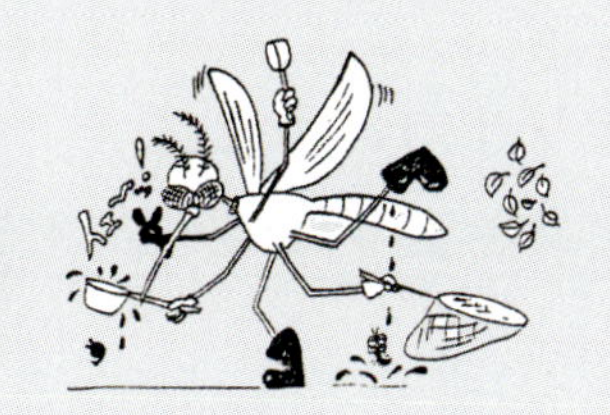

【蚊帳とのかかわり】

その3. 殺虫剤浸透蚊帳の効果

WHO（世界保健機構）は 1992 年頃から殺虫剤浸透性蚊帳（低毒性殺虫剤ピレスロイドなどを予め蚊帳の繊維に組み入れた特殊加工した蚊帳）によるマラリア防圧対策をアフリカのマラリア流行地で実施し，幼児の死亡率の激減に好成績を得た．

ラオスのカムアン県でも 1999 年殺虫剤浸透性蚊帳配布（日本国無償援助）によるマラリアの疫学的調査が計画され，私も媒介蚊の生態調査に参加した．

ラオスの中南部に位置するカムアン県のボラパ各地区の住民の蚊帳配布前（1996，1997 年）のマラリア陽性率は 22 ～ 54%であった．殺虫剤浸透性蚊帳配布を終えた 1999 年，特にオリセットネット（殺虫剤ピレスロイド浸透蚊帳）を 配布した村での陽性率は急激に減少（8 ～ 22%）した．媒介蚊は *Anopheles dirus* と *Anopheles minimus* で経産率（長生きをしている蚊の割合）も蚊帳配布後は激減した（宮城・當間，2017）．

現在，ボラパ地区はマラリアの流行はなく，風光明媚な農村地になっていると聞いた．（宮城）

蚊は家の中に壁の隙間から自由に侵入

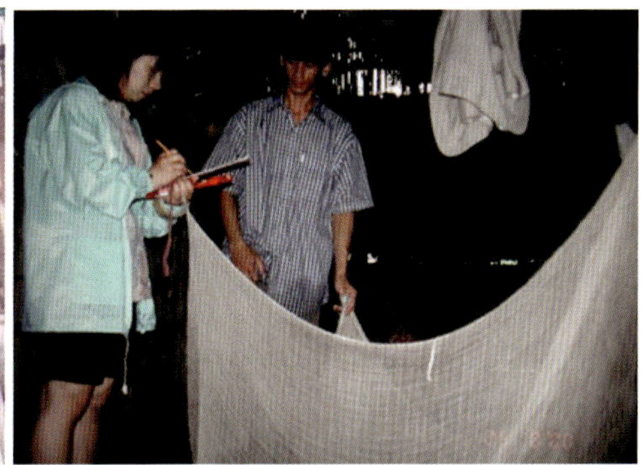

蚊帳をチェックする隊員

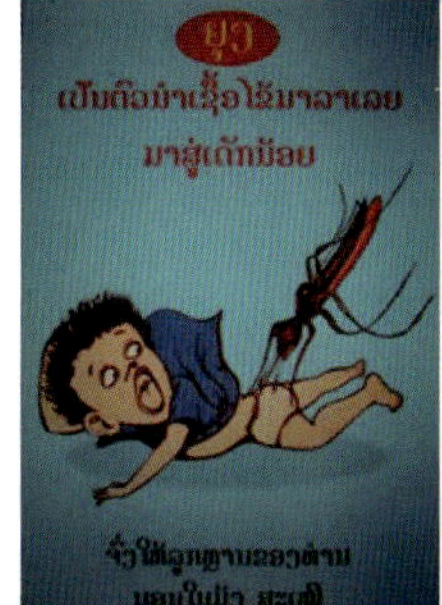

マラリア予防に関するポスターが目につく

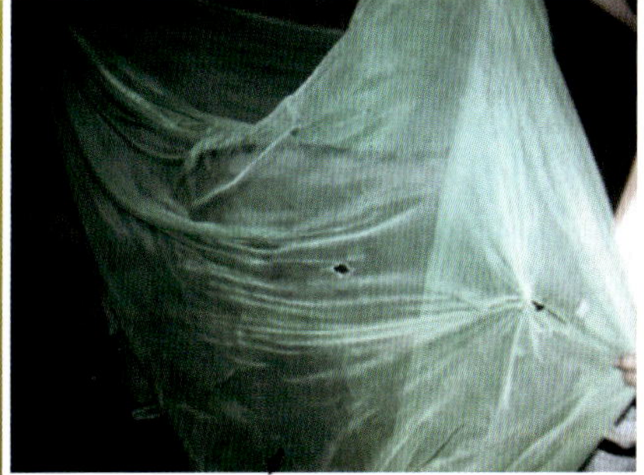

穴が開いた蚊帳を使用

1999 年ラオスのカムアン県での殺虫剤浸透性蚊帳配布によるマラリアの疫学的調査

20. カエルの鳴き声に誘引されるチスイケヨソイカとチビカ

動画 20J

上映時間
3 分 33 秒

コレスレラ属（*Corethrella*）はケヨソイカ科（Chaoboridae）の1亜科として長い間取り扱われてきた．しかし蚊によく似た翅脈や吸血性口器を持つことなどケヨソイカの他の亜科の種とは著しく異なることから独立したチスイケヨソイカ科（Corethrellidae）が提唱された．チスイケヨソイカは，蚊科やケヨソイカ科の姉妹群（Sister group）と位置づけられている．チスイケヨソイカは1属約63種と化石から発見された2種を含む小さなグループで，ほとんどの種が中南米から記録されている．我が国からは4種が記録されている；ヤマトチスイケヨソイカ *C. japonica*，トワダチスイケヨソイカ *C. towadensis*，ニホンチスイケヨソイカ *C. nippon*，リュウキュウチスイケヨソイカ *C. urumense*．幼虫は湿地性で捕食性と思われるが生態は不明である．雌成虫はもっぱらカエルの血を吸うことからカエル吸血性コバエ Frog biting midges と呼ばれ，カエルの鳴き声に誘引される．カエル鳴き声トラップを考案し，西表島の湿地帯で採集をした結果，2種のチスイケヨソイカとマクファレンチビカが多数採集された．これらは両生類の住血寄生虫の媒介者と思われる．

Corethrella **biting midges and** ***Uranotaenia*** **mosquitoes attracted to frog calls**

Video 20E

Show time
3m 46 s

The genus *Corethrella* has long been treated as a subfamily of the family Chaoboridae. However, because it is remarkably different from other subfamilies of the Chaoboridae, such as having wing veins and blood-sucking mouthparts that closely resemble those of Culicidae mosquitoes, a separate family Corethrellidae, has been proposed. It is positioned as a sister group of Culicidae. The family is a small group consisting of about 63 species in one genus (*Corethrella*) and two fossil species, most of which have been recorded from Central and South America. Four species have been recorded from Japan; *C. japonica, C. towadensis*, *C. nippon* and *C. urumense*. The larvae are found in swampy areas; they have predatory habits but their biology is unknown. The adults exclusively feed on the blood of frogs, and they are attracted to the sound of frogs. These *Corethrella* species in Okinawa are potential carriers of blood protozoan parasites, *Trypanosoma* and Haemogregarines detected in amphibians and reptiles.

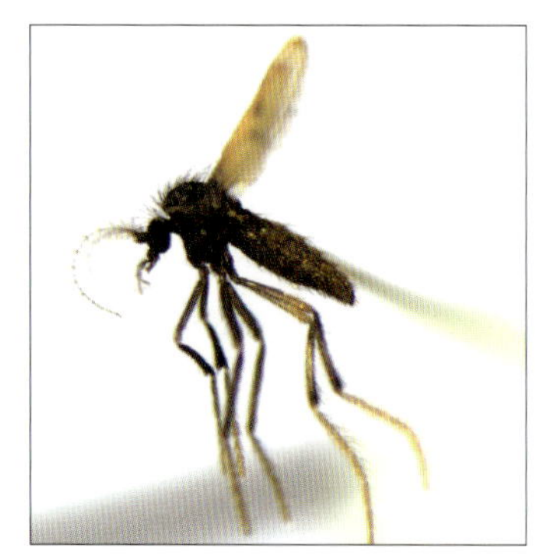

リュウキュウチスイケヨソイ雌
C. urumense adult female

フタクロホシチビカ雌
Ur. n. novobscura adult female

カエルを吸血する蚊の吸血行動は
The blood-sucking behavior of mosquito that fed on frog:

カエル鳴き声トラップ：カエルの鳴き声を CD プレーヤーで拡大，飛来した昆虫を吸引器で捕集する
Frog calling trap: Frog calling is amplified on a CD player and flying insects are captured with a suction device

●蚊の起源　（比嘉由紀子）

蚊は小さく，体は硬い外殻に覆われているわけではない．加えて，幼虫の発生源である水溜まりは海や湖沼のような何万年も存在する広大な水域に比べるとはるかに小さく，ほとんどが短期間で消滅してしまう．そのような事情から，蚊の化石は現在まで世界中からわずか数十個しか発見されていない．ほとんどが新生代以降の化石で，中生代と考えられる化石は3個，これまで，現存する化石の中で最も古い蚊とされていたのはミャンマーの中生代白亜紀中期の地層の琥珀の中から発見された *Burmaculex antiquus* Borkent and Grimaldi（Borkent and Grimaldi, 2004）と考えられていた．この琥珀の推定年代は9000万年～1億年である．そんな中，2023年に世界を驚かせた研究成果が発表された．中生代白亜紀前期のレバノンの琥珀に閉じ込められている蚊を最新の技術を用いて形態を精査した結果，①これまでのミャンマーの最古の記録よりも3000万年さらにさかのぼると推定された．②近縁科のケヨソイカ科の雄や現在の蚊の雄に比べてレバノンで発見された雄の吻が長く，雌のように歯状の吸血に適した吻の構造を持っていた．③蚊科の新種と考えられ，新亜科（Libanoculicinae subfam. nov.），新属（*Libanoculex* gen. nov.），新種（*Libanoculex intermedius* sp. nov.）として記載されたが，すでに絶滅していると考えられた（Azar et al., 2023）．*Libanoculex intermedius* と同時期に蚊の起源を考察するうえで他にも重要な研究が相次いで発表されている．蚊に最も近縁なケヨソイカ科昆虫の古い化石から蚊の起源は少なくとも推定1億8000万年前にさかのぼると考えられていたが（Borkent and Grimaldi, 2004），中生代三畳紀後期から中期（2億4720万年～2億130万年前）と推定されるさらに古いケヨソイカの化石が見つかっている（Lukashevich, 2022）．また，蚊科の102種のDNAを用いた研究で，現在の主要な蚊科のグループであるハマダラカ亜科とナミカ亜科は中生代ジュラ紀前期頃（1億9750万年前）に分岐したと推定されている（Lorenz et al., 2021）．これらの結果から，ケヨソイカからの蚊の分岐年代（起源）がこれまで考えられていたころよりもさらに古い可能性がでてきた．前述したように，蚊のように微細な昆虫の化石は残りにくく，化石から推定される事象についてまだまだ埋められないギャップは多い．しかし，解析技術と研究の進歩で蚊の起源が明らかになる日も決して遠いことではないかもしれない．

さて，最近見つかった雄が吸血していた可能性のある *Libanoculex intermedius* に話を戻す．現在，科学的な一般常識として，蚊は卵の発育，産卵のために雌のみが動物から吸血すると考えられている．雌が無吸血で産卵する種や生命活動のためにシリアゲアリが集めた栄養分をカギカ属の雄がもらう事例はわずかながら確認されているが，世界中から記録されている約3,700種の蚊科の昆虫の中で，雄が動物から吸血することはまだ確認されていない．これらの事実から，雄が吸血していた可能性を示す化石が発見されたことはとても大きな発見である．*Libanoculex intermedius* の雄の小顎肢を調べたところ，現代の蚊と同様の空気の流れや二酸化炭素を感じる感覚器が存在することが明らかになっている．吻の構造や感覚器の存在が，*Libanoculex intermedius* の雄が吸血していたと考えられている理由である．雌が産卵のために血液を必要とすることを考えると，*Libanoculex intermedius* の雄は吸血することでとてもアクティブに行動できたと考えられるが，そのような有利な習性を現代の蚊の雄がなぜ失ってしまったかは，残念ながらわかっていない（Azar et al, 2023）．

Origin of mosquitoes

Mosquitoes are tiny, and their bodies are fragile. In addition, their larval habitats are not the vast water bodies such as oceans and lakes that have existed for thousands of years but small water bodies, most of which disappear quickly. For these reasons, only a few dozen mosquito fossils have been discovered around the world to date. Most of them are from the Cenozoic era or later, but there are three fossils that are thought to be from the Mesozoic era. Until now, the oldest mosquito fossil in existence is the *Burmaculex antiquus* Borkent and Grimaldi (Borkent and Grimaldi, 2004), discovered in amber from the mid-Cretaceous period of the Mesozoic era in Myanmar. The estimated age of this amber is 90 to 100 million years.

Meanwhile, in 2023, one research surprisingly reported that using the latest technology, the morphology of a mosquito preserved in Lebanese amber from the Early Cretaceous period of the Mesozoic era was 1) found to be 30 million years older than the oldest known record in Myanmar, 2) the male found in Lebanon had a more extended proboscis than the male of the Culicidae at present and closely related family, the Chaoboridae, and its structure of proboscis is suitable for blood-sucking like a female, and 3) it was described as a new subfamily (Libanoculicinae subfam. nov.), a new genus (*Libanoculex* gen. nov.), and a new species (*Libanoculex intermedius* sp. nov.), but the species was already extinct (Azar et al, 2023).

Other critical studies on the origin of mosquitoes have been published at the same time as *Libanoculex intermedius*. Based on old fossils of the family Chaoboridae, which are most closely related to mosquitoes, it was thought that the origin of mosquitoes dates back at least 180 million years ago (Borkent and Grimaldi, 2004), but even older Chaoboridae fossils estimated to be from the late to middle Triassic period of the Mesozoic era (247.2 to 201.3 million years ago) have been found (Lukashevich, 2022). In addition, a study using DNA from 102 species of the family Culicidae has estimated that the current major groups of mosquitoes, the Anophelinae and Culicinae subfamily, diverged around the Early Jurassic period of the Mesozoic era (197.5 million years ago) (Lorenz et al., 2021). These results suggest that the divergence (origin) of mosquitoes from Chaoboridae may be even older than previously considered. As mentioned above, fossils of tiny insects such as mosquitoes are challenging to preserve, and there are still many gaps to be filled regarding events inferred from fossils. However, with the advancement of technology and research, the day when the origin of mosquitoes will be revealed may not be far.

Now, let's go back to the recent topic of *Libanoculex intermedius*, whose males may have been sucking blood. Currently, it is scientifically common knowledge that only female mosquitoes suck blood from animals to develop and lay eggs. There are a few species in which females lay eggs without sucking blood, and males of the genus *Malaya* obtain nutrients collected by ants for their vital activities. However, among the approximately 3,700 species of mosquitoes recorded from the world, no males suck blood from animals. Given these facts, the discovery of a fossil indicating the possibility that males may have been sucking blood was a massive discovery. When the maxillae of male *Libanoculex intermedius* were examined, it was revealed that they had sensilla that sense for carbon dioxide, like those of modern mosquitoes. The structure of the proboscis and the presence of sensilla are the reasons why those males of *Libanoculex intermedius* sucked blood. Given that females need blood to lay eggs, it is likely that males of *Libanoculex intermedius* could have been very active by sucking blood, but unfortunately, it is not known why modern male mosquitoes have lost this advantageous behavior (Azar et al., 2023). (Y. Higa)

21. 蟻の口から栄養源（食物）を頂戴するカギカの生態
―パプアニューギニアでのカギカ *Malaya leei* との出会い―

動画 21J

上映時間
5 分 59 秒

カギカ属は小さなグループで 12 種が熱帯アフリカ，アジア，オーストラリアなどから記録されている．成虫はギンモンカ属の蚊とよく似て，背中に銀色の縦線がみられるが，口吻の先端が膨らみ，長い毛や鱗片でおおわれており，ほかの属と容易に区別される．この蚊は動物の血や花の蜜などを吸う習性はなく，もっぱらシリアゲアリの口から栄養液を吸い取ることにより，生きながらえ，増殖している．カギカの口吻の先端部の長い毛が蟻の口に触れると，蟻はたまりかねて，中腸内の液体を吐き戻すと思われている．カギカはこの吐き出された糖分を含む液体を満腹になるまで繰り返しなめるようにして吸い上げる．パプアニューギニアで遭遇したカギカ（*Malaya leei*）の摂食行動を動画で示した (Miyagi, 1981)．また，インドネシアのカギカ属の蚊（*Malaya genurostris*）がシリアゲアリから摂食している写真を紹介した．このような蟻に依存する生物を好蟻性生物 “Myrmecophiles” と呼んでいる．カギカ属の幼虫は種々の草本植物の葉腋の水溜まりに発生し，水底の堆積物や壁に固着する菌類を口刷毛でかき集めて摂食している．

Feeding habits of myrmecophilous *Malaya leei* encountered at Papua New Guinea

Video 21E

Show time
5 m 51 s

The genus *Malaya* is a small group of mosquitoes: 12 species are represented by the Afrotropical, Australasia and Oriental Regions. They are small mosquitoes, ornamented with brilliant silvery, blue or violet scales. They are closely related to the genera *Topomyia* and *Kimia* having a silvery line on the center of the scutum, but are distinguished from all genera of Culicidae by their distinctive hairy and clubbed proboscis. The feeding behavior of *Malaya* is unique. The fly alighted in front of the advancing ant and held the ant by its front legs. When the hairy proboscis of the fly touched the ant's mouth, the ant might feel tickled and regurgitated the food in its stomach. The fly is sucking exclusively the regurgitated fluid. Without this fluid, adult *Malaya* mosquito cannot survive (Miyagi, 1981). The mosquito is a “Myrmecophile”. The larvae are leaf axil breeder, found commonly in Taro plants in Okinawa. The feeding mode of the larva is “Collecting-filtering” and “Collecting-gathering“, having well developed mouth-brushes with the middle ones at least serrated distally.

蟻の口から栄養液を奪い取るカギカ．
小松貴氏　撮影

The fly steals nutrient fluid from an ant’s mouth.
Photo by Mr. T. Komatsu

オキナワカギカ雌成虫（正面と側面撮影）

Malaya genurostris female (front and lateral facing views)

クワズイモの群落
Colony of Taro plant *Alocasia*

スポイドで葉腋の水溜まりに生息する幼虫を採集

Collect larvae living in water puddles in leaf axils using a pipette

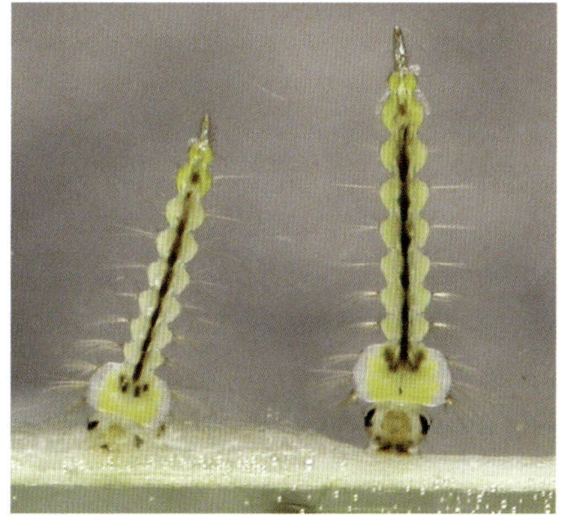

葉腋の溜まり水の底の堆積物を摂食する幼虫

Larvae that feed on accumulated sediment at the leaf axils of Taro plant

22. いろいろな動物の血を吸う蚊 ― 蚊の吸血嗜好性 ―

動画 22J

上映時間
7分18秒

蚊は雄雌ともに花の蜜を吸うが，多くの蚊は雌だけが産卵（繁殖）のために動物の血を吸う．また，蚊は吸血する動物をある程度選んでいる．哺乳類や鳥類などの温血動物だけではなく，両生・爬虫類，魚（トビハゼ）などの冷血動物を好んで吸血する蚊も意外に多い．また，全く血を吸わない蚊もいる．最も特殊化したカギカ属の蚊は栄養物をもっぱら蟻の口から奪い取ることによって生命を維持し，繁殖している．これまでにフィールドや実験室内で観察・撮影した蚊の吸血シーンをまとめて編集した．写真を提供していただいた各位に深く感謝する．

Food preference of mosquitoes: Blood-feeding female mosquitoes obtain their blood-meals from a wide variety of hosts

Video 22E

Show time
7m 17 s

The adult mosquitoes, both male and female are nectar feeders. Only the females feed on blood. However, not all species have blood-sucking females. The blood-sucking female mosquitoes obtain their blood-meals from a wide variety of animals, including both the cold-blooded and warm-blooded vertebrates. The mosquitoes with a special preference for human are called "Anthropohilic", while those restricted their feeding to animals, both wild and domestic, are "Zoophilic". In *Toxorhynchites* and *Topomyia* mosquitoes, blood-sucking habit is absent; they are "Autogenous". The extraordinary food habits of mosquito are found in the genus *Malaya,* which feed on regurgitated stomach contents offered by *Crematogaster* ants.

森林内に生息する種々の蚊に吸血されているイリオモテヤマネコ．奥土晴夫氏　撮影
Iriomote wildcat being sucked by various mosquitoes living in the forest. Photo by Mr. H. Okudo

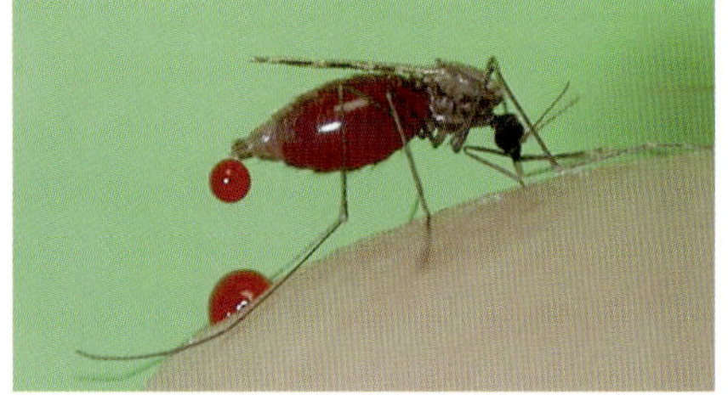

人の血を吸うコガタハマダラカ
An. yaeyamaensis fed on human

吸血習性の進化　Evolution of blood-sucking habits

●琉球列島の蚊の吸血源動物（當間孝子）

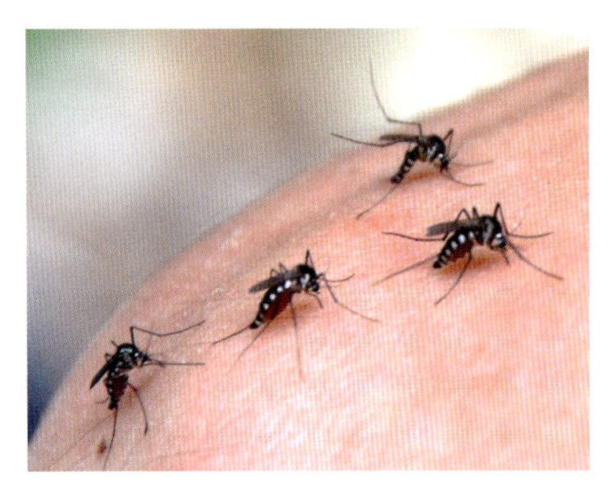

ヒトの血を吸うアマミムナゲカ
Hz. kana feeding on human

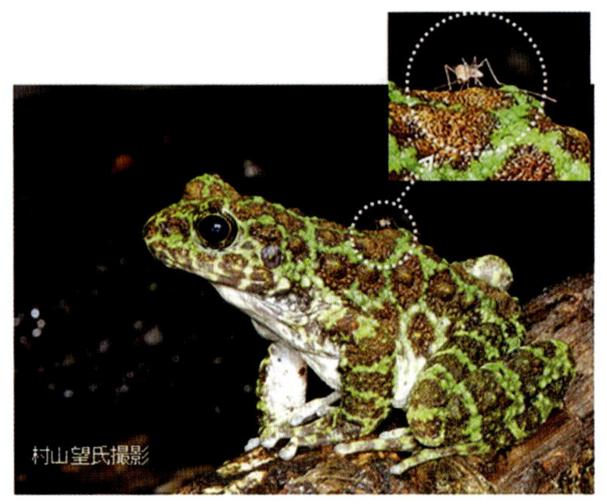

イシカワガエルを吸血するチビカ（村山望氏 撮影）
Uranotaenia sp. feeding on *Odorrana ishikawae*
Photo by Mr. N. Murayama

最近，アメリア合衆国カルフォルニア州のフロリダに生息しているチビカ属の *Uranotaenia sapphirina* (Osten Sacken) が環形動物（ミミズ，ヒル）を吸血していることが報告されている（Reeves et al., 2018）．脊椎動物だけでなく，無脊椎動物を吸血する蚊の発見は，蚊がこれまで認識されていた範囲より広範囲の動物分類群を利用していることを示している．蚊は雌成虫のみが産卵のために，吸血を行う．無吸血で産卵する種もいる．蚊が自然界でどんな動物から吸血しているのか，吸血習性や嗜好性を明らかにすることは，蚊が媒介する病気の予防対策を行う上で大切なことである．最近は吸血性ハエ目（Diptera）の進化学的な研究が注目されている（Azar et al., 2023; Borkent and Grimaldi, 2004; Lorenz et al., 2021）．

蚊の吸血嗜好性を明らかにする方法として，1）吸血している現場を突き止める（直視法），2）野外で，いろいろな動物を囮にして吸血飛来する蚊を集める（動物囮法），3）実験室でケージの中に成熟した雌蚊といろいろな動物を同棲させ，吸血行動を観察する（見合い法）（宮城，1972），4）自然界で吸血直後の蚊を採集し，蚊の中腸（胃）内の血液の DNA を抽出，増幅し，分析後，検索するなどの方法（PCR 法）がある（宮城・當間，2017）．宮城・當間（2017）が述べているように，いずれの方法も一長一短があるが，これらの方法を駆使して自然界における各種蚊の吸血習性を模索しなければならない．

DNA による吸血源動物の特定の方法は最近，盛んに行われるようになっている．この方法は正確ですぐれているが，設備と試薬に費用が掛かり，吸血直後の蚊を多数採集することが容易でない．Tamashiro et al.（2011）は，琉球列島の蚊の吸血嗜好性を知るために，種々の方法で吸血蚊を採集し，DNA を分析する方法で，11 属 35 種 975 個体の吸血蚊の吸血源動物を明らかにしている．同様な方法で，Sawabe et al.（2010）は人家周辺に生息する蚊，Toma et al.（2014）はチビカ *Uranotaenia* 属の蚊 7 種の吸血源動物を同定している．その後，Toma et al.（2024）は，7 種（ヤブカ *Aedes* 属 2 種，フトオヤブカ *Verrallina* 属 1 種，イエカ *Culex* 属 4 種）について吸血源を明らかにした．Miyake et al.（2019）は琉球列島各地のオキナワアナジャコの巣穴に休息するカニアナヤブカの腹部内の血液を分析し，各地で異なる魚類を同定している．最近は，蚊と病原体の関係を明らかにする方法として，蚊から分離された病原体のDNAを検出する研究も急増している．その例として鳥マラリアの研究がある (Ejiri et al., 2008, 2009, 2011；Kim et al., 2009；津田 , 2017, 2019).

本著では琉球列島に生息の記録がある全77種について，これまで報告された結果に新知見（當間・宮城，未発表）を加えて付表4に示した．本付表では吸血源動物は温血動物，冷血動物に別け，さらに温血動物をヒト，哺乳類，鳥類に，冷血動物を爬虫類，両生類，魚類に分け，それぞれの該当動物分類群欄に○で示した．蚊が人の病気との関わりについて今後考察が行いやすいようにするために，ヒトと他の哺乳類を吸血している場合やヒトのみを吸血している場合は，ヒトと哺乳類の両方に○を，ヒト以外の哺乳類を吸血している場合は，哺乳類のみに○を書き入れた．吸血源の動物群が初めて明らかになった場合は◎で示した．琉球列島では生息数が少なく直視法が困難な種，吸血個体を採集することができなかった種や琉球列島では絶滅したと思われる種については，他県やアジアの近隣諸国での結果を示した．情報が見つからない種は，吸血源動物を考える上で参考になる近縁亜種の吸血源について言及した．各種蚊の吸血嗜好性を検索するに際しては付表4だけでなく，同時に付表1～3も見ていただくとより理解しやすい．以下にこれまでに明らかになった各種蚊の吸血源動物についての知見を詳しく示す.

モンナシハマダラカは琉球列島が分布の北限地で，奄美大島と徳之島に生息し，個体数が少なく，吸血習性や吸血源は不明である．本種はマレーシアでは，ヒトを吸血している（Reid, 1968).

オオツルハマダラカはヒト吸血嗜好性が高く，ウシ，ブタ，イノシシ，ヤギやスイギュウの大動物も吸血する（Ho, 1965；Tamashiro et al., 2011).

ヤマトハマダラカは日本や韓国ではヒトよりウシに誘引される（Reid, 1968).

オオハマハマダラカは昼間，沖縄島北部の林内に入ると直ちにヒトに吸血に飛来し，イノシシも吸血している（Toma and Miyagi, 1986；Tamashiro et al., 2011；Mannen et al., 2016).

シナハマダラカはヒト，イヌ，ウシ，ブタ，ヤギ，スイギュウやウマを吸血する（鳥羽，1919；Bohart and Ingram, 1946；Toma and Miyagi, 1986；Tamashiro et al., 2011，當間・宮城，未発表)，実験室ではマウスやヒヨコを吸血する（宮城，1972).

タテンハマダラカの発生は少なく，成虫はライトトラップやドライアイストラップで採集されているが，吸血蚊は採集されていない．東南アジアではヒトやウシを吸血している（Reid, 1968).

コガタハマダラカは強い人吸血嗜好性が見られている（宮城・当間，1978)．ウシも吸血する（Tamashiro et al., 2011).

キンイロヤブカは畜舎にライトトラップを設置すると多くの吸血蚊を得ることができる．ヒト，ウシ，ブタ，ニワトリやヌマガエルを吸血し，沖縄島の動物園内ではキリンやサイも吸血している（Tamashiro et al., 2011；Sawabe et al., 2010)．実験室ではマウス, ヒヨコ, カエルと共にヘビも吸血する（宮城，1972).

オキナワヤブカはヒト，イノシシやクビワオオコオモリを吸血する（Tamashiro et al., 2011; Bohart and Ingram, 1946).

ヤエヤマヤブカの発生数は少ないが，西表島の森林内でヒト（宮城・当間，1980）に吸血飛来する .

ニシカワヤブカはトカラ列島の中之島でヒトに吸血に飛来した（Toma and Miyagi, 1986）.

カニアナヤブカはサキシマヌマガエルや魚類のハゼを吸血し（Okudo et al., 2004；Tamashiro et al., 2011），西表島ではジャノメハゼ，沖縄島と奄美大島ではゴマホタテウミヘビが主要な吸血源である（Miyake et al., 2019）.

ムネシロヤブカは希な種で，西表島古見の森林内で，昼間，1個体の雌がヒトに吸血に飛来したのを採集したのみである（Toma and Miyagi, 1986）.

アマミヤブカは奄美大島，徳之島特産種で，ヒトに吸血に飛来した（Tamashiro et al., 2011）.

サキシマヤブカは石垣島と西表島に生息するが，採集個体数が少なく，吸血習性については不明である．九州以北に生息する亜種のヤマトヤブカは，昼間吸血性でヒトを吸血し，畜舎でも吸血個体が採れる（津田, 2019）．実験室ではマウスやヒヨコを吸血する（宮城，1972）.

ナンヨウヤブカは，琉球列島では希な種で，成虫のみがライトトラップで採集されている（Toma and Miyagi, 1986）が，吸血源動物は不明である．シンガポールではヒト，ウシやスイギュウを吸血する（Colless, 1959）.

ハマベヤブカは黒島の海岸のグランドプールで幼虫が一度採集されているのみである（Toma and Miyagi, 1986）．ニューカレドニアではフィラリアの媒介蚊として知られ，人畜を激しく攻撃する．ヒト，イヌ，フクロギツネ，ネコ，オオコオモリや鳥類が吸血源動物として記録されている（Wilkerson et al., 2021）.

ワタセヤブカは沖縄島の森林内でヒトに吸血に飛来した（福地ら, 2021）．本種は，昼間吸血性で，ヒトを吸血する（津田, 2019）.

ネッタイシマカは琉球列島では最近採集されず（Higa et al., 2007），1970年石垣島の川平でヒトに吸血に飛来したのが最後の記録である（Tanaka et al., 1975）.

ヒトスジシマカは，野外でヒト，ジャコウネズミやウサギを吸血する（Tamashiro et al., 2011）が，新たにネコ，ウシ，ブタやニワトリも吸血していることが明らかになった（當間・宮城，未発表）．Ejiri et al.（2008）は南大東島の本種から鳥マラリアを検出している．実験室内ではヘビ，カメやカエルも吸血する（宮城，1972）.

ダイトウシマカは南・北大東島にのみ生息し，神社境内の林内に入ると直ちにヒトに吸血に飛来する（Toma and Miyagi, 1986）.

ダウンスシマカは森林内でヒトに吸血に飛来する（Bohart and Ingram, 1946；Toma and Miyagi, 1986）.

ミヤラシマカは石垣・西表島の森林内でヒト（Tamashiro et al., 2011）やヤエヤマセマルハコガメ（Toma et al., 2024）を吸血する.

リバースシマカはヒト，ウシ，クビワオオコオモリ，リュウキュウヤマガメを吸血し（Tamashiro et al., 2011），ヤギやイノシシも吸血している（當間・宮城，未発表）.

トウゴウヤブカは，野外でヒトやトリを吸血し（Toma and Miyagi, 1986；中村ら, 1987；Wilkerson, 2021），実験室内ではヘビ，カメやカエルも吸血する（宮城，1972）．血を吸わないで少数の卵を産む個体がいる（Mogi, 1981）.

オオクロヤブカは人家周辺で発生し，ヒト，ウシ，ブタ，ヤギ，ガチョウ，ダチョウ，沖縄島の動物園ではキリン，サイ（Tamashiro et al., 2011）やニワトリ（當間・宮城，未発表）を吸血し，実験室内では低率ではあるがヘビを吸血する（宮城，1972）.

アマミムナゲカは奄美大島と徳之島の特産種である．林内に立ち入ると直ちに吸血に飛来する人吸血嗜好性の高い蚊である

(Tamashiro et al., 2011).

フトオヤブカ属の 3 種は人吸血嗜好性が高く，**コガタフトオヤブカ**は他に爬虫類のヤエヤマセマルハコガメを，**アカフトオヤブカ**はイノシシを吸血する他，今回初めて爬虫類（ヤエヤマセマルハコガメ）を吸血していた．**クロフトオヤブカ**はヒトを激しく襲う（Tamashiro et al., 2011；Toma et al., 2024；當間・宮城，未発表）.

オビナシイエカは野外でヒト（Colless, 1959）やウシ（Tamashiro et al., 2011）を吸血する .

ジャクソンイエカは生息個体数が少なく，吸血源は不明である．

ミナミハマダライエカはヒト（Toma and Miyagi, 1986）やイノシシ（Tamashiro et al., 2011）を吸血する．

シロハシイエカはウシ，ブタ，ニワトリ（Tamashiro et al., 2011; 當間・宮城，未発表）を吸血し，シンガポールではヒト，ウシ，ブタ，トリ，イヌやウマを吸血していることが報告されている（Colless, 1959）.

ネッタイイエカはヒト，ネコ，イヌ，ニワトリ，ダチョウ，ノドグロツグミ，ヌマガエル（Tamashiro et al., 2011），ウシ，クビワオオコオモリ（Sawabe et al., 2010），ブタ，ヤギやヒヨドリを吸血する他，今回，初めて爬虫類（オンナダケヤモリ）を吸血していた（當間・宮城，未発表）.

ヨツホシイエカはシンガポールでヒト，ウシ，ブタ，イヌやトリを吸血している（Colless, 1959）.

コガタアカイエカはヒト，ウシ，ブタ，ヤギ，スイギュウ，イノシシやニワトリ，沖縄島の動物園ではキリンやサイを吸血し（Tamashiro et al., 2011），実験室ではマウスやヒヨコと共にヘビを吸血する（宮城，1972）.

スジアシイエカはヒトを吸血し（Sirivanakarn, 1976），兵庫県の野外で採集された個体はノドグロツグミを吸血していた（Toma et al., 2024）.

ニセシロハシイエカはウシ，ニワトリやダチョウ，沖縄島の動物園ではサイを吸血している個体が採集されている（Tamashiro et al., 2011）．台湾では本種（*Cx. anurus* として発表）のみが日本脳炎ウイルスの陽性蚊であり，重要な媒介蚊であると結論づけ，東南アジアでは重要な日本脳炎の媒介蚊である（Sirivanakarn, 1976）.

セシロイエカはヒトや大型哺乳動物を吸血し（津田，2019），バンクロフト糸状虫の中間宿主である（Sirivanakarn, 1976）.

クロフクシヒゲカはゴイサギ（Toma et al., 2024）を吸血し，石垣島で採集された本種からは鳥マラリア原虫が検出されている（Ejiri et al., 2011）．シンガポールではウシ，トリを吸血している個体が採集されている（Colless, 1959）.

アカクシヒゲカは実験室でヒヨコは吸うが，マウス，ヘビやカエルは吸わないことが報告されている（宮城，1972）．本種からバンクロフト糸状虫の幼虫が分離されているが，実際には媒介にあまり関与してない（Bram, 1967）.

リュウキュウクシヒゲカはヒト，イノシシやアオカナヘビを吸血する（Tamashiro et al., 2011）.

カギヒゲクロウスカは発生個体数が少なく，吸血習性は不明である．東南アジアではヒトや家畜などは吸血しないことが報告されている（Sirivanakarn, 1972）.

リュウキュウクロウスカはナミエガエル，ホルストガエルやヤエヤマアオガエル（Tamashiro, 2011；當間・宮城，未発表）を吸血している．宮山（2020）は，本種がオットンガエルを吸血している写真撮影に成功している．

オキナワクロウスカはオットンガエルを吸血している（Toma et al., 2024）.

クロツノフサカ，**カニアナツノフサカ**の吸血習性は不明である．

ハラオビツノフサカはシンガポールではイヌ，ウマやトリを吸血する（Colless, 1959）．

フトシマツノフサカはウシやサキシマヌマガエル（Tamashiro et al., 2011；Toma et al., 2024）を，実験室ではマウス，ヒヨコ，ヘビ，カメやカエルも吸血する（宮城，1972）．

アカツノフサカは実験室でヒキガエルを吸血する（山田，1932）．

カラツイエカはヒト，ウシ，ニワトリ（Toma and Miyagi, 1986; Tamashiro et al., 2011）やヒヨドリ（當間・宮城，未発表）を，島根県出雲平野で採集した個体はヒトやダイサギを吸血していた（津田，2013）．

ミツホシイエカはヒトやウマを吸血する（Bohart and Ingram, 1946；Toma and Miyagi, 1986）．

サキジロカクイカは南大東島で採集された個体から鳥マラリア原虫が検出されている（Ejiri et al., 2008）．

トラフカクイカは西表島で採集された個体がシジュウカラを吸血し（Tamashiro et al., 2011），神奈川県の動物園や東京港野鳥公園で採集された個体からは鳥マラリア原虫が検出されている（Ejiri et al., 2009；Kim et al., 2009）．島根県出雲平野で採集された個体はニホンカナヘビを吸血していた（津田，2013）．

オキナワエセコブハシカはイノシシを吸血する（Tamashiro et al., 2011）．

マダラコブハシカはリュウキュウカジカガエル（Tamashiro et al., 2011）や，ヤエヤマアオガエルを吸血する（當間・宮城，未発表）．

ルソンコブハシカはウシだけでなく，リュウキュウカジカガエル，シロアゴガエル（Tamashiro et al., 2011），ヌマガエルを吸血する（當間・宮城，未発表）．

沖縄島の林内で，昼間，**ムラサキヌマカ**はヒトに吸血に飛来した（福地ら，2021）．本種は，昼間，人畜を激しく吸血することが報告されている（佐々ら，1976）．石垣・西表島で行った人囮（二重蚊帳）や牛囮蚊帳法で本種は多数採集されている（宮城・当間，1978，1980）．マレーシアでは，本種は鳥嗜好性であることが報告されている（Delfinado, 1966）．

キンイロヌマカは人畜を吸血し，石垣島では人囮（二重蚊帳）や牛囮蚊帳法で採集されている．ムラサキヌマカに比べると誘引数は少ない（佐々ら，1976；宮城・当間，1978）．

アシマダラヌマカはウシやヤギの大動物吸血嗜好性が高い（Tamashiro et al., 2011）．石垣島の森林内や沖縄島の具志頭海岸では，しばしばヒトに吸血に飛来している（宮城・当間，1978；中村ら，1987）．2018年，沖縄島のうるま市照間の湿地で，昼間本種に激しく襲われた（宮城・當間，未発表）．シンガポールではヒト，ウシやトリを吸血している（Colless, 1959）．

ハマダラナガスネカは幼虫が沖縄島北部の森林内の樹洞や人工容器には多数生息しているが，成虫はヒトの血を吸いに飛来しない（Toma and Miyagi, 1986）．宮城（1972）は，実験室でヒヨコを吸血するが，マウス，ヘビ，カメやカエルを吸血しないことを報告している．

オキナワカギカは動物を吸血せず，シリアゲアリから餌（栄養液）を得て繁殖している（Barraud, 1934；宮城・當間，2017）．本種の分布は琉球列島が北限で，幼虫はクワズイモの葉腋の水溜まりに普通に生息しているが，成虫が蟻の口から餌を得ている場面に出くわした報告はない．Miyagi（1981）はパプアニューギニアで本種と近縁の *Malaya leei* (Wharton) が蟻から餌を得ている光景に遭遇し，8ミリフイルムに収めている．

ヤンバルギンモンカは吸血せずに，産卵を行い，世代を繰り返している（Toma and Miyagi, 1986；Okazawa et al., 1986：宮城・當間，2017）．

ヤエヤマナガハシカは石垣島の林内でしばしば吸血に飛来する（宮城・当間，1978）．与那国島ではヤエヤマイシガメを吸血している写真が撮影されている（宮城・當間，2017）．実験室内ではマウスを用いて累代飼育を行うことができる（当間・宮城，1983）．旧北区日本（屋久島，種子島以北）に生息している同属のキンパラナガハシカは実験室内でマウス，ヒヨコ，ヘビやカメを吸血し，カエルは吸血しない（宮城，1972）．本種の室内実験で，幼虫期に高温・長日条件から低温・短日条件に移すと，少数の雌が吸血せずに，卵を産む．すなわち無吸血産卵を行うことが明らかになっている（森章夫，1976）．

オオカ属の成虫雌は吸血せず，花の蜜や，樹液を吸って生活をしている（宮城・當間，2017）．

カニアナチビカはナミエガエル，オキナワアオガエル，リュウキュウカジカガエル，クロイワトカゲモドキ（Tamashiro et al., 2011；Toma et al., 2014）やオットンガエルを吸血している（當間・宮城，未発表）．

ムネシロチビカはアイフィンガーガエル（Tamashiro et al., 2011；Toma et al., 2014）を吸血している．

リュウキュウクロホシチビカはカエルを吸血する（Sawabe et al., 2010）が，カエル鳴き声トラップ（Toma et al., 2005），ライトトラップ（Toma et al., 2007）やヒト（宮城・当間，1980）には誘引されない．島根県出雲平野で採集された近縁亜種のフタクロホシチビカはヒト，ウシやシュレーゲルアオガエルを吸血している（津田，2013）．実験室のお見合い法ではカエルを吸血するが，カメ，ヘビ，ヒヨコやマウスは吸血しない（宮城，1972）．

シロオビカニアナチビカはヤエヤマハラブチガエル，オオハナサキガエル，サキシマヌマガエル，アイフィンガーガエル，ヤエヤマアオガエルやヒメアマガエルなどの多くのカエルを吸血し，さらにハゼの仲間を吸血する（Tamashiro et al., 2011；Toma et al., 2014）．

イリオモテチビカの吸血源動物は不明である．

ハラグロカニアナチビカはサキシマヌマガエル，ヤエヤマアオガエルやヒメアマガエルを吸血している（Tamashiro et al., 2011; Toma et al., 2014）．

オキナワチビカはナミエガエル，オキナワアオガエルやサキシマヌマガエルを吸血している（Tamashiro et al., 2011; Toma et al., 2014）．

コガタチビカは西表島ではサキシマヌマガエル（Toma et al., 2014），シンガポールではウシ（Colless, 1959），マレーシアでは魚類（Tempelis, 1975）を吸血している．

マクファレンチビカはナミエガエル，オキナワアオガエル，リュウキュウカジカガエル，ヤエヤマハラブチガエル，サキシマヌマガエル，アイフィンガーガエルやヒメアマガエルなどの多くのカエルを吸血している（Tamashiro et al., 2011；Toma et al., 2014）．

今後，多くの吸血蚊が採集され，分析が進むことによって，野外での本来の種々の蚊の吸血習性や吸血嗜好性がさらに明らかになることを期待する．

Blood-sucking preferences of mosquitoes in the Ryukyu Archipelago

Recently, it has been reported that *Uranotaenia sapphirina* (Osten Sacken), which lives in Florida, California in the United States, feeds on annelids (earthworms and leeches) (Reeves et al., 2018). The discovery of mosquitoes that feed on invertebrates as well as vertebrates indicates that mosquitoes utilize a wider range of animal taxa than previously recognized. Only adult female mosquitoes feed on blood for oviposition. Some species can lay eggs without feeding on blood. It is important to clarify the blood-sucking habits and preferences of mosquitoes for implementing preventive measures against mosquito-borne diseases. Methods to clarify mosquito blood-feeding preferences include: (1) identifying the site of blood-feeding (Direct visualization method); (2) using various animals as bait to collect mosquitoes in the field (Animal baited net trap method); (3) housing mature female mosquitoes with various animals in cages and observing their blood-feeding behavior in laboratories (Matchmaking method) (Miyagi, 1972); and (4) collecting mosquitoes immediately after feeding on blood in the field, extracting, amplifying, analyzing and searching for host DNA from the blood in the mosquito's midgut (PCR method). As Miyagi and Toma (2017) stated, each method has its advantages and disadvantages, but these methods must be utilized to explore the blood-sucking habits of various mosquitoes in the field. The DNA-based methods for identifying the blood-meal sources have recently become popular. Although this method is accurate and excellent, the equipment and reagents are expensive, and it is difficult to collect large number of blood-fed mosquitoes in the field. To understand the blood-sucking preferences of mosquitoes in the Ryukyu Archipelago, Tamashiro et al. (2011) collected blood-sucking mosquitoes using various methods and analyzed the DNA to identify the blood-meal sources for 975 blood-sucking mosquitoes representing 35 species belonging to 11 genera. Using a similar method, Sawabe et al. (2010) identified the blood-meal sources for mosquitoes collected at the residential areas, and Toma et al. (2014) also identified the blood-meal sources for seven species of *Uranotaenia* mosquitoes. Subsequently, Toma et al. (2024) identified the blood-meal sources for seven mosquito species (two *Aedes*, one *Verrallina*, and four *Culex* species). Miyake et al. (2019) analyzed the blood in the midguts of crab hole *Aedes* mosquitoes resting in the burrows of Okinawan burrowing crabs in main islands of the Ryukyu Archipelago and identified different fish species in each island. Recently, there has been a rapid increase in research into detecting the DNA of pathogens isolated from mosquitoes as a method to clarify the relationship between mosquitoes and pathogens, such as research into avian malaria (Ejiri et al., 2008, 2009, 2011; Kim et al., 2009; Tsuda, 2017, 2019).

In this study, all 77 mosquito species recorded in the Ryukyu Archipelago are listed in Appendix 4, with previously reported results and new findings (Toma and Miyagi, unpublished). In this appendix, blood-providing animals are divided into warm-blooded and cold-blooded animals, with warm-blooded animals further divided into humans, mammals, and birds, and cold-blooded animals into reptiles, amphibians, and fish, and each is indicated with a circle in the corresponding animal taxonomic group column. In order to make it easier to consider the relationship between mosquitoes and mosquito-borne human diseases in the future, if a mosquito feeds on both humans and other mammals, or only on humans, we mark both humans and mammals with a circle, and if a mosquito feeds on a mammal other than humans, we mark only the mammal. When the animal group that is the source of blood identified for the first time, it is indicated with ◎. For the species with small populations in the Ryukyu Archipelago, species for which direct observation is difficult, species for which blood-sucking specimens could not be collected, and species thought to be extinct, results from other prefectures and neighboring Asian countries are presented. For the species where no information is available, the blood-meal sources of closely related subspecies are used as reference when considering blood-meal sources. When searching for the blood-sucking preferences of various mosquitoes, it will be easier to understand if you refer to not only Appendix 4 but also Appendices 1−3 at the same time. Below is a detailed description of the blood-meal sources for the mosquito species. Recently, evolutionary study on hematophagous flies (Diptera) has attracted attention (Azar et al., 2023; Borkent and Grimaldi, 2004; Lorenz et al., 2021).

Anopheles **(*Anopheles*)** ***bengalensis***: The northernmost distribution limit of this species is the Ryukyu Archipelago, Amamioshima and Tokunoshima. The population of this species is small, and its blood-sucking habits and blood source are unknown. In Malaysia, this species feeds on human blood (Reid, 1968).

An. **(*Ano.*)** ***lesteri***: It has been reported that this species has a high preference for human blood feeding (Ho, 1965), although it also feeds on cattle, *Bos taurus*, pig and wild boar, *Sus scrofa*, goat, *Capra hircus*, and water buffalo, *Bubalus arnee* (Ho, 1965; Tamashiro et al., 2011).

An. **(*Ano.*)** ***lindesayi japonicus***: In Korea, it is more attracted to cattle than to human (Reid, 1968).

An. **(*Ano.*)** ***saperoi***: During the day, this species immediately flies to feed on human in the forests of the northern part of Okinawajima. It feeds on wild boar (Toma and Miyagi, 1986; Tamashiro et al., 2011; Mannen et al., 2016).

An. **(*Ano.*)** ***sinensis***: It feeds on the blood of human, dog, *Canis lupus familiaris*, cattle, goat, pig, water buffalo, and horse, *Equus caballus* (Toba, 1919; Bohart and Ingram, 1946; Toma and Miyagi, 1986; Tamashiro et al., 2011; Toma and Miyagi, unpublished), and in the laboratory, it feeds on mouse, *Mus musculus*, and fowl, *Gallus gallus* (Miyagi, 1972).

An. **(*Cellia*)** ***tessellatus***: The occurrence of this mosquito is low, and although adult mosquitoes have been collected using light traps and dry ice traps, no blood-fed mosquitoes have been collected in the field. In Southeast Asia, they feed easily on human and cattle (Reid, 1968).

An. **(*Cel.*)** ***yaeyamaensis***: This mosquito has been shown to have a strong preference for human blood (Miyagi and Toma, 1978). It also sucks blood from cattle (Tamashiro et al., 2011).

Aedes **(*Aedimorphus*)** ***vexans nipponii***: This mosquito can be caught in large numbers by light traps in livestock pens. The females feed on the blood of human, cattle, pig, fowl, and Indian rice frog, *Fejervorya kawamurai.* At a zoo in Okinawajima, it has even been found to feed on giraffe, *Giraffa camelopardalis* and white rhinoceros, *Ceratotherium simum* (Sawabe et al., 2010; Tamashiro et al., 2011). In the laboratories, it feeds on snake as well as mouse, fowl, and frog (Miyagi, 1972).

Ae. **(*Bruceharrisonius*)** ***okinawanus okinawanus:*** It feeds on the blood of human, wild boar, and Ryukyu flying fox, *Pteropus dasymallus* (Bohart and Ingram, 1946; Tamashiro et al., 2011).

Ae.* (*Brh.*) *o. taiwanus: Although the number of the mosquitoes is low, they fly to the forests of Iriomotejima to suck blood from human (Miyagi and Toma, 1980).

Ae.* (*Downsiomyia*) *nishikawai: This mosquito feeds on human blood at Nakanoshima (Toma and Miyagi, 1986).

Ae.* (*Geoskusea*) *baisasi: This mosquito feeds on Sakishima rice frogs, *Fejervarya sakishimensis*, and fish, goby (Okudo et al., 2004; Tamashiro et al., 2011), and its main source of blood is four-eyed sleeper, *Bostrychus sinensis*, on Iriomotejima, and fish, rice-paddy eel, *Pisodonophis boro* on Okinawajima and Amamioshima (Miyake et al., 2019).

Ae.* (*Hopkinsius*) *albocinctus: It is a rare species, and only one female was collected during the daytime in the forest of Komi, Iriomotejima, when it flew to feed on human blood (Toma and Miyagi, 1986).

Ae.* (*Hulecoeteomyia*) *j. amamiensis: This mosquito flies to humans to suck blood (Tamashiro et al, 2011).

***Ae.* (*Hul.*) *j. yaeyamensi*s**: This mosquito inhabits Ishigakijima and Iriomotejima. The number of individuals collected is small, and its blood-sucking habits are unknown. *Aedes japonicus* that inhabits Kyushu and the north, is a daytime blood-sucking species that feeds on human, and blood-fed females were found in livestock barns (Tsuda, 2019). It has been reported to feed on mouse and fowl in the laboratories (Miyagi, 1972).

Ae.* (*Neomelaniconion*) *lineatopennis: This species is rare in the Ryukyu Archipelago, and only adults have been collected by light trap (Toma and Miyagi, 1986). Its blood source is unknown. In Singapore, it feeds on human, cattle and water buffalo (Colless, 1959).

Ae.* (*Ochleotatus*) *vigilax: Larvae of this species have only been collected once, from a ground pool on the coast of Kuroshima, Yaeyama Islands (Toma and Miyagi, 1986). In New Caledonia, it is a vector of filariasis and aggressively attacks human. The documented hosts include human, dog, cat, *Felis catus*, brushtail possum, flying fox and birds (Wilkerson et al., 2021).

Ae.* (*Phagomyia*) *watasei: The mosquito came to feed on human in the forests of Okinawajima (Fukuchi et al., 2021). It is a diurnal blood-feeding species that feeds on human (Tsuda, 2019).

Ae.* (*Stegomyia*) *aegypti: It has not been recently collected in the Ryukyu Archipelago (Higa et al., 2007); the last record was in 1970 when it came to feed on human blood at Kabira, Ishigakijima (Tanaka et al., 1975).

Ae.* (*Stg.*) *albopictus: In the field, this species feeds on human, shrew, *Suncus murinus* and rabbit, *Oryctolagus cuniculus* (Tamashiro et al., 2011). It has recently been reported to feed on cat, cattle, pig, and fowl (Toma and Miyagi, unpublished). Ejiri et al. (2008) detected avian malaria in this species on Minamidaitojima. In the laboratories, it also feeds on snake, turtle, and frog (Miyagi, 1972).

Ae.* (*Stg.*) *daitensis: This mosquito is an endemic species in Minamidaitojima and Kitadaitojima. The female immediately flies to human to suck blood within the shrine grounds in the forest (Toma and Miyagi, 1986).

Ae.* (*Stg.*) *f. downsi: This mosquito flies to human near its habitat to feed on human blood (Bohart and Ingram, 1946; Toma and Miyagi, 1986).

Ae.* (*Stg.*) *flavopictus miyarai: The mosquito feeds on human (Tamashiro et al., 2011) and tortoise, yellow-margined box turtle, *Cuora flavomarginata evelynae* (Toma et al., 2024) blood.

Ae.* (*Stg.*) *riversi: This mosquito feeds on the blood of human, cattle, Ryukyu flying fox and tortoise, Japanese black-breasted leaf turtle, *Geoemyda japonica* (Tamashiro et al., 2011), as well as goats and wild boars (Toma and Miyagi, unpublished).

Ae.* (*Tanakaius*) *togoi: This mosquito feeds on human and bird in the field (Toma and Miyagi, 1986; Nakamura et al., 1987; Wilkerson, 2021), and on snake, turtle, and frog in laboratories (Miyagi, 1972). Some females lay small numbers of eggs without feeding on blood (Mogi, 1981).

Armigeres subalbatus: This mosquito feeds on blood from humans, cattle, pigs, goats, goose, *Anser anse*r and ostrich, *Struthio camelus* as well as giraffes and white rhinoceros at a zoo in Okinawajima (Tamashiro et al., 2011), and fowl (Toma and Miyagi, unpublished); it feeds on snake at a low rate in the laboratories (Miyagi, 1972).

Heizmannia* (*Heizmannia*) *kana: This species is endemic to Amamioshima and Tokunoshima, and is highly selective in feeding on human blood, flying immediately upon entering forests to feed on human blood (Tamashiro et al., 2011).

Verrallina* (*Harbachius*) *nobukonis: In northern Okinawajima and Iriomotejima, the larvae of this mosquito sometimes appear in large numbers in the forests near river mouths and fly to human to suck blood. This mosquito feeds on yellow-margined box turtle (Toma et al., 2024).

Ve.* (*Neomacleaya*) *atriisimilis: This mosquito feeds on human and wild boar (Tamashiro et al., 2011; Toma and Miyagi, unpublished), and has been found to feed on the blood of reptile, yellow-margined box turtle, for the first time (Toma and Miyagi, unpublished).

Ve.* (*Verrallina*) *iriomotensisis: This mosquito feeds readily on human (Tamashiro et al., 2011; Toma et al., 2024; Toma and Miyagi, unpublished).

Culex* (*Culex*) *fuscocephala: This mosquito feeds on blood from human (Colless, 1959) and cattle (Tamashiro et al., 2011) in the field.

Cx.* (*Cux.*) *jacksoni: This mosquito is rare in the Ryukyu Archipelago and its blood-sucking habits are unknown.

Cx.* (*Cux.*) *mimeticus: This mosquito feeds on human (Toma and Miyagi, 1986) and wild boar (Tamashiro et al., 2011).

Cx.* (*Cux.*) *pseudovishnui: This mosquito feeds on cattle, pig, and fowl (Tamashiro et al., 2011; Toma and Miyagi, unpublished), and has been reported in Singapore to feed on human, cattle, pig, fowl, dog, and horse (Colless, 1959).

Cx.* (*Cux.*) *quinquefasciatus: This mosquito feeds on the blood of human, cat, dog, fowl, ostrich, bird, dark-throated thrush, *Turdus ruficollis*, Indian rice frog (Tamashiro et al., 2011), cattle, Ryukyu flying fox (Sawabe et al., 2010), pig, goat, and bird, brown-eared bulbul, *Hypsipetes amaurotis*, and has been found to feed on the blood of reptile, four-clawed gecko, *Gehyra mutilata*, for the first time (Toma and Miyagi, unpublished).

Cx.* (*Cux.*) *sitiens: This mosquito feeds on human, cattle, pig, dog and fowl in Singapore (Colless, 1959).

Cx.* (*Cux.*) *tritaeniorhynchus: This mosquito feeds on the blood of human, cattle, pig, goat, water buffalo, wild boar, fowl, and giraffe and white rhinoceros at a zoo in Okinawajima (Tamashiro et al., 2011). In the laboratories, it feeds on snake as well as mouse and fowl (Miyagi, 1972).

Cx.* (*Cux.*) *vagans: This mosquito feeds on humans (Sirivanakarn, 1976), and one female collected in Hyogo Prefecture was feeding on the blood of dark-throated thrush (Toma et al., 2024).

Cx.* (*Cux.*) *vishnui: This species has been found feeding on cattle, fowl, and ostrich, and white rhinoceros at a zoo in Okinawajima (Tamashiro et al., 2011). In Taiwan, this species (published as *Cx. anurus*) is the only mosquito found to be positive for Japanese encephalitis virus, and it is considered to be an important vector for Japanese encephalitis in Southeast Asia (Sirivanakarn, 1976).

Cx.* (*Cux.*) *whitmorei: This mosquito feeds no blood from humans and large mammals (Tsuda, 2019) and is an intermediate host for *Wuchereria bancrofti* (Sirivanakarn, 1976).

Cx.* (*Culiciomyia*) *nigropunctatus: This mosquito feeds on bird, black-crowned night heron, *Nycticorax nycticorax* (Toma et al., 2024) and avian malaria parasites have been detected in this species collected on Iriomoteijima (Ejiri et al., 2011). In Singapore, this species has been found feeding on cattle and fowl (Colless, 1959).

Cx.* (*Cul.*) *pallidothorax: It has been reported that this mosquito feeds on fowl, but not on mouse, snake or frog in the laboratory (Miyagi, 1972). Human filaria (*Wuchereria bancrofti*) larvae have been isolated from this species, but they are not actually involved in transmission (Bram, 1967).

Cx.* (*Cul.*) *ryukyensis: This mosquito feeds on the blood of human, wild boar, and reptile, green grass lizard, *Takydromus smaragdinus* (Tamashiro et al., 2011).

Cx.* (*Eumelanomyia*) *brevipalpis: The population of this mosquito is small, and its blood-sucking habits are unknown. In Southeast Asia, it has been reported that it does not suck blood from humans or livestock (Sirivanakarn, 1972).

Cx.* (*Eum.*) *h. ryukyuanus: This mosquito feeds on the blood of Namie's frog, *Limnonectes namiyei*, Holst's frog, *Babina holsti*, and Owston's green tree frog, *Rhacophorus owstoni*, (Tamashiro, 2011; Toma and Miyagi, unpublished). Miyama (2020) successfully photographed this species feeding on the blood of an Otton frog, *Babina subaspera*.

Cx.* (*Eum.*) *okinawae: This mosquito feeds blood from Otton frog (Toma et al., 2024).

Cx.* (*Lophoceraomyia*) *bicornutus: The blood-sucking habits of this mosquito are unknown.

Cx.* (*Lop.*) *cinctellus: In Singapore, this mosquito sucks the blood of dog, horse and fowl (Colless, 1959).

Cx.* (*Lop.*) *infantulus: This mosquito feeds on cattle and Sakishima rice frogs (Tamashiro et al., 2011; Toma et al., 2024), and in the laboratory it also feeds on mouse, fowls, snakes, turtles and frogs (Miyagi, 1972).

Cx.* (*Lop.*) *rubithoracis: This mosquito feeds on the blood of Japanese common toads (*Bufo japonicus japonicus*) in the laboratories (Yamada, 1932).

Cx.* (*Lop.*) *tuberis: The blood-sucking habits of this mosquito are unknown.

Cx.* (*Oculeomyia*) *bitaeniorhynchus: This mosquito has been found to feed on human, cattle, fowl (Toma and Miyagi, 1986; Tamashiro et al., 2011), and bird, brown-eared bulbuls (Toma and Miyagi, unpublished); the mosquito collected from Izumo in Shimane Prefecture had fed on human and bird, great egret, *Ardea alba* (Tsuda, 2013).

Cx.* (*Ocu.*) *sinensis: This mosquito feeds on the blood of human and horse (Bohart and Ingram, 1946; Toma and Miyagi, 1986).

Lutzia* (*Metalutzia*) *fuscana: Avian malaria parasites have been detected in the mosquito collected on Minamidaitojima (Ejiri et al., 2008).

Lt.* (*Mlt.*) *vorax: This mosquito collected on Iriomotejima feeds on Japanese tit, *Parus major* (Tamashiro et al., 2011), and avian malaria parasites have been detected in specimens collected at a zoo in Kanagawa Prefecture and Tokyo Port Wild Bird Park (Ejiri et al., 2009; Kim et al., 2009). A specimen collected in the Izumo Plain in Shimane Prefecture feeds on Japanese grass lizard, *Takydromus tachydromoides* (Tsuda, 2013).

Ficalbia ichiromiyagii: This mosquito feeds on the blood of wild boars (Tamashiro et al., 2011).

Mimomyia* (*Etorleptiomyia*) *elegans: This mosquito sucks the blood of Ryukyu Kajika frog, *Buergeria japonica*, (Tamashiro et al., 2011) and Owston's green tree frog (Toma and Miyagi, unpublished).

Mi.* (*Eto.*) *luzonensis: This mosquito not only sucks blood from cattle, but also from Ryukyu Kajika frog, white-chinned tree frog, *Polypedates leucomystax* (Tamashiro et al., 2011), and Indian rice frog (Toma and Miyagi, unpublished).

Coquillettidia* (*Coquillettidia*) *crassipes: This mosquito flew to suck blood (Fukuchi et al., 2021). It has been reported to suck blood from human and livestock vigorously during the day (Sasa et al., 1976). Large numbers of this species have been collected using human bait (Double mosquito net) and cattle bait net methods on Ishigakijima and Iriomotejima (Miyagi and Toma, 1978, 1980). In Malaysia, this species has been reported to be ornithophilous (Delfinado, 1966).

Cq.* (*Coq.*) *ochracea: This mosquito feeds on the blood of humans and livestock, and on Ishigakijima it has been collected using human (Double mosquito nets) and cattle bait net methods. The number of this mosquito is smaller than that of *Cq. crassipes* in the Ryukyu Arcipelago (Sasa et al., 1976; Miyagi and Toma, 1978).

Mansonia* (*Mansonioides*) *uniformis: This mosquito has a high preference for feeding on large animals such as cattle and goat (Tamashiro et al., 2011). In the forests of Ishigakijima and Gushikami coast on Okinawajima, it often comes to feed on human (Miyagi and Toma, 1978; Nakamura et al., 1987). In 2018, we were violently attacked by this species during the day in the wetlands of Teruma, Uruma City, Okinawajima (Miyagi and Toma, unpublished). In Singapore, it feeds on human, cattle and fowl (Colless, 1959).

Orthopodomyia anopheloides: The larvae of this mosquito are found in large numbers in tree cavities and artificial containers in the forests of the northern part of Okinawajima, but the adults do not fly to humans to suck blood (Toma and Miyagi, 1986). Miyagi (1972) reported that in the laboratory, they suck blood from fowls, but do not suck blood from mouse, snake, tortoise, or frog.

Malaya genurostris: This mosquito does not suck blood from animals, but reproduces by obtaining food (nutritional fluid) from the ants, *Crematogaster* sp. (Barraud, 1934; Miyagi and Toma, 2017). The distribution of this species is at the northern limit of the Ryukyu Archipelago, and the larvae commonly inhabit pools of water in the axils of Taro plants, but there have been no reports of adults obtaining food from the mouths of ants. Miyagi (1981) encountered a closely related species, *Mala-*

ya leei (Wharton), in Papua New Guinea obtaining food from ants, and captured the scene on 8 mm film.

Topomyia* (*Suaymyia*) *yanbarensis: This mosquito lays eggs without feeding any blood (Toma and Miyagi, 1986; Okazawa et al., 1986: Miyagi and Toma, 2017).

Tripteroides* (*Tripteroides*) *yaeyamensis: This mosquito often flies to human to feed on blood in the forests of Ishigakijima (Miyagi and Toma, 1978). On Yonagunijima, it has been photographed sucking the blood of tortoise, an Asian brown pond turtle, *Mauremys mutica kami* (Miyagi and Toma, 2017). It is possible to breed the mosquito using mouse blood in the laboratory for successive generations (Toma and Miyagi, 1983). *Tripteroides bambusa* which lives in the Palearctic Region of Japan, sucks blood from mouse, fowl, snake, and tortoise in laboratories, but does not suck blood from frog (Miyagi, 1972). Laboratory experiments with this species have revealed that when the larval stage is transferred from high temperature and long day conditions to low temperature and short day conditions, a small number of females lay eggs without sucking blood (Mori, 1976).

Toxorhynchites species –***Toxorhynchites* (*Toxorhynchites*) *manicatus yaeyamae*, *Tx.* (*Tox.*) *m. yamadai*** and ***Tx.* (*Tox.*) *okinawensis***: Adult females of the genus do not suck any blood, but instead survive by sucking nectar from flowers (Miyagi and Toma, 2017).

Uranotaenia* (*Pseudoficalbia*) *jacksoni: This mosquito feeds on the blood of Namie's frog, Okinawa green tree frog, *Rhacophorus viridis viridis*, Ryukyu Kajika frog, Japanese ground gecko, *Goniurosaurus kuroiwae kuroiwae* (Tamashiro et al., 2011; Toma et al., 2014), and Otton frog (Toma and Miyagi, unpublished).

Ur.* (*Pfc.*) *nivipleura: This mosquito feeds on the blood of Eiffinger frog, *Kurixalus eiffingeri* (Tamashiro et al., 2011; Toma et al., 2014).

Ur.* (*Pfc.*) *novobscura ryukyuana: This mosquito will suck the blood of frog (Sawabe et al., 2010), but is not attracted to CDC miniature frog call traps (Toma et al., 2005), light traps (Toma et al., 2007), or human baited traps (Miyagi and Toma, 1980). The closely related subspecies, ***Ur.* (*Pfc.*) *novobscura novobscura***, collected in the Izumo Plain, Shimane Prefecture, sucks the blood of human, cattle, and Schlegel's green tree frog, *Rhacophorus schlegelii* (Tsuda, 2013). In matchmaking method in the laboratories, the mosquito will suck the blood of frog, but not tortoise, snake, fowl and mouse (Miyagi, 1972).

Ur.* (*Pfc.*) *ohamai: This mosquito feeds on the blood of many frogs, including the Yaeyama spotted frog, *Rana okinavana*, giant nosed frog, *Odorrana supranarina*, Sakishima rice frog, Eifinger frog, Owston's green tree frog, and small yellow-spotted frog, *Microhyla okinavensis*, as well as gobies (Tamashiro et al., 2011; Toma et al., 2014).

Ur.* (*Pfc.*) *tanakai: The source animal of blood for this species is unknown.

Ur.* (*Pfc.*) *yaeyamana: This mosquito feeds on the blood of Sakishima rice frog, Owston's green tree frog, and small yellow-spotted frog (Tamashiro et al., 2011; Toma et al., 2014).

Ur.* (*Uranotaenia*) *annandalei: This mosquito feeds on the blood of Namie's frog, Okinawa green tree frog, and Sakishima rice frog (Tamashiro et al., 2011; Toma et al., 2014).

Ur.* (*Ura.*) *lateralis: This mosquito feeds on the blood of Sakishima rice frog on Iriomotejima (Toma et al., 2014), cattle in Singapore (Colless, 1959), and fish in Malaysia (Tempelis, 1975).

Ur.* (*Ura.*) *macfarlanei: This mosquito feeds on the blood of many kinds of frogs, including Namie's frog, Okinawa green tree frog, Ryukyu Kajika frog, Yaeyama spotted frog, Sakishima rice frog, Eiffinger frog, and small yellow-spotted frog (Tamashiro et al., 2011; Toma et al., 2014). (T. Toma)

23. 蚊幼虫の基本的な摂食法：収集濾過法と収集かじり取り法

動画 23J

上映時間
7 分 40 秒

Collecting-filtering（収集濾過法）：幼虫は水表面に垂れ下がり，呼吸しながら口刷毛で水流を起こし，水中や水表面を浮遊する微粒子を口元に引き寄せ体内に取り込む．例外として，ハマダラカは水面で，ヌマカは水中で摂食する．イエカ，ハマダラカ，ヌマカ，コブハシカ，ヤブカの一部がこの方法で摂食している．

Collecting-gathering（収集かじり取り法）：水底の堆積物を口刷毛でかき混ぜて小動物の死骸や藻などをあさり，かみくだいて摂食する．多くのヤブカ，ナガハシカ，チビカ属がこの方法で摂食している．ヤンバルギンモンカはかじり取り法（Scraping），カラツイエカは裁断法（Shredding）として区別される．

The feeding modes commonly used by mosquito larvae: Collecting-filtering mode and Collecting-gathering mode

Video 23E

Show time
6 m 53 s

Collecting is the feeding mode commonly used by the larvae. Collecting-filtering : Removal of food particles suspended in the water column or associated with the water surface. It is found characteristically among species of *Anopheles, Culex, Mansonia, Ficalbia* and some other genera, and less commonly in *Aedes* (Clements, 1992).

Collecting-gathering mode: Removal of particles deposited on or in submerged materials and loosely attached to surfaces. The mode may be defined as the process of abrasion of solid materials, and needing further manipulation by mouthpart before entering the digestive tract. It is found in many species of *Aedes*, *Tripteroides* and *Uranotaenia*.

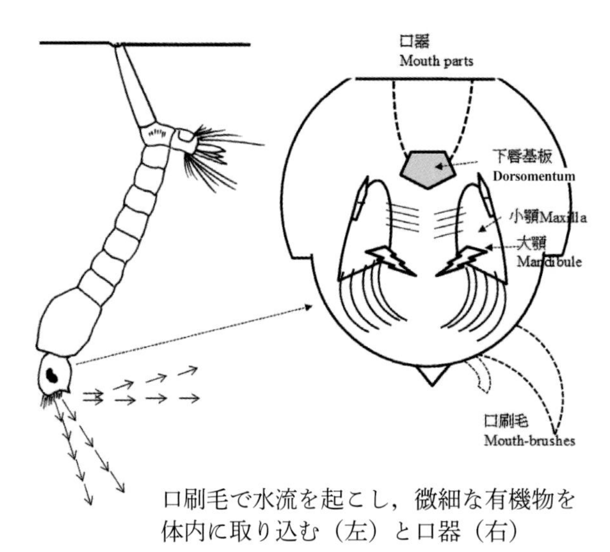

口刷毛で水流を起こし，微細な有機物を体内に取り込む（左）と口器（右）

The larvae use their mouth-brushes to current (left) and take tiny organic matter into their mouth parts (right)

幼虫の基本的な摂食方法 Typical feeding modes of larvae

摂食法 Feeding mode	**口刷毛** Mouth-brusheS	**該当する蚊** Examples	
収集濾過法 Collecting-filtering	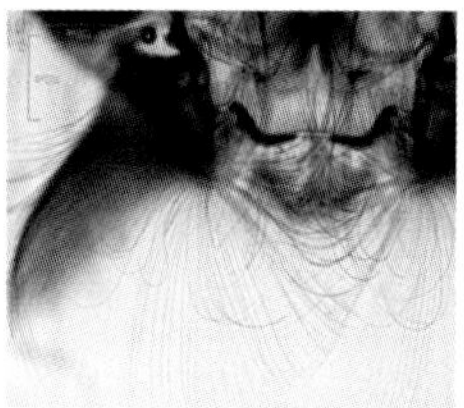	 ネッタイイエカ *Cx. quinquefasciatus*	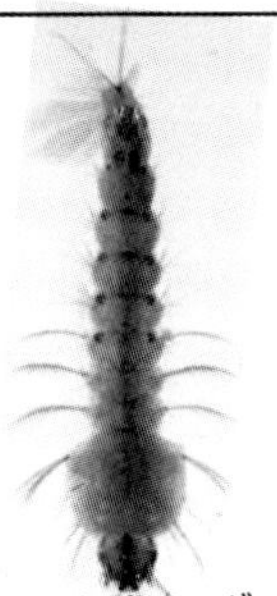 シナハマダラカ *An. sinensis*
収集かじり取り法 Collecting-gathering	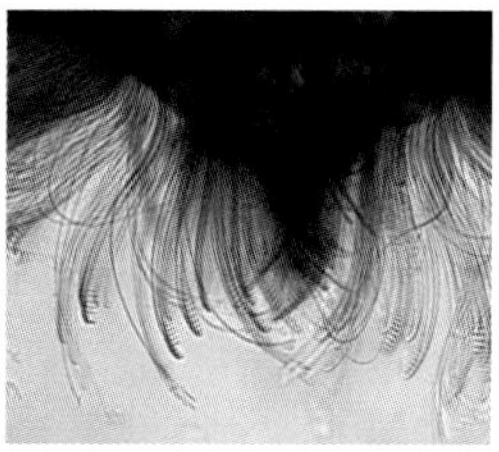	 ヒトスジシマカ *Ae. albopictus*	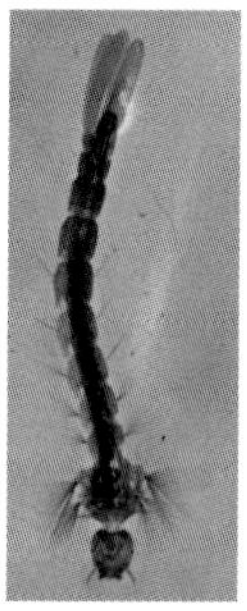 オオクロヤブカ *Ar. subalbatus*
捕食 Predation	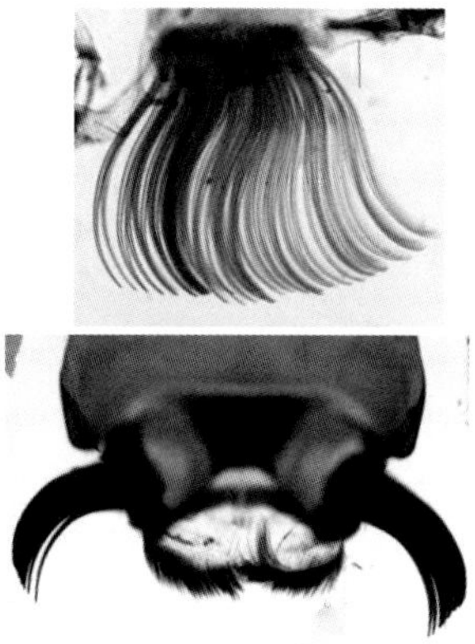	 カクイカ *Lutzia*	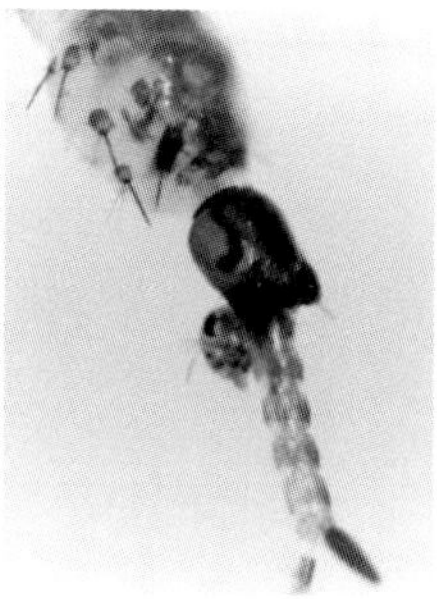 オオカ *Toxorhynchites*

●ボウフラの採餌行動観察の面白さ，有用さ（宮城一郎）

蚊の幼虫を普通ボウフラと呼んでいる．体を棒のようにくねくね曲げて水中を遊泳することから「棒振り」と呼ばれ，後にボウフラと呼ばれるようになったとされている．庭のかたすみの陰地に放置され，数日前の大雨で雨水が溜まった空き缶に，いつの間にか多数のボウフラが発生した．これらのボウフラを溜水と水底の堆積物と一緒に持ち帰り，数個の小さい水槽（巾 10 cm × 高さ 7 cm × 奥行 2.5 cm）に移し入れ，いつも休憩するソファーの横に置いてお茶を飲みながらボウフラの動きを観察し，撮影をはじめた．また，近くの水田や湿地の水溜まりで多数のボウフラを採集した．丁度コロナ禍でステイホームが始まったころから 3 年間の日課になった．

全てのボウフラ（ヌマカ類を除く）は水面に呼吸管（Siphon）の先端（Spiracle）を出して空中の酸素を体内に取り入れて呼吸している．そのためボウフラの動きや摂食は呼吸と深く関係している．ほとんどの種のボウフラは雑食性で，餌は水表面や水中を漂う微細な有機物（動・植物由来）か，水底の堆積物（昆虫など小動物の死骸）である．水槽を泳ぐボウフラを眺めていると，遊泳には主に二つの方法があることがわかる．一つは体をくねらせ，頭部を後にして水の表面と水底を移動する「くねくね移動」（Side to side lashing movement）ともう一つは体をくねらせることなく，頭を先に「横滑り移動」（Gliding movement）である．「くねくね移動」は主に振動などで驚いた時にみられる．「横滑り移動」は頭の先に生えているひと房のしなやかな口刷毛（Mouth-brushes または Lateral palatal brush）をなびかせて頭を先にして水面を移動している．また，水面で静止（呼吸）している姿勢，遊泳動作，採餌動作は個体によって多少異なり，属（Genus）や種（Species）の特徴となっている．

庭の空き缶のボウフラ 40 個体はいずれも 3，4 齢であった．これらの幼虫の水中での遊泳（動き），水面での静止姿勢，餌の食べ方，形態（大きさ，色）など肉眼で 3 つのグループ A, B, C（種類）に区別された．

A グループは 22 個体，頭幅約 1 cm，呼吸管は細長く，口刷毛が発達し，水面で静止し，時には横滑りして口刷毛を振動させ，水中の浮遊物を口元に引き寄せて食べている（収集濾過法，Collecting-filtering mode）．水面に垂れ下がり一斉に口刷毛を動かしている様子はかわいらしい．

B グループは 15 個体，頭幅は A と変わらないが，呼吸管は A に比較して太く短い．水面と水底をくねくね上下移動し，水底の堆積物に頭部を突込み物色し，有機物をかじり取っている．ときには堆積物の小さな塊をくわえて水面に浮上し，塊を風車の様に回転させ摂食している（収集かじり取り法，Collecting-gathering mode）．この様子は街角の大道芸人のようである．

C グループは 3 個体，前 2 グループに比べて明らかに大きく，頭部はおむすび型（三角），頭幅約 1 cm，呼吸管は太くて短い．動きは少なく，接近する幼虫にかみつき食べている（捕食性 Predation）．過日アフリカのサバンナで見たライオンがガセルを捕食している姿を連想した．

水田や湿地で採集したボウフラも前 3 グループと異なる 4 グループ D，E，F，G に分けることができた．D は A に頭幅，動き，採餌方法などよく似ているが，呼吸管が明らかに長い．E は呼吸管が長く形態は A，D に似ている．幼虫はいつも緑藻の塊の中で緑藻の糸を食べている．そのために幼虫の腹部内の腸管は緑色を帯びている．幼虫はアオミドロの長い細糸を触角（Antenna）と口刷毛でたぐり寄せ，大顎（Mandible），小顎（Maxilla），下唇基板(Dorsomentum)で糸を2分する．一方の糸断片を頭部に留置きし，もう一方の断片を大顎で裁断（Shredding）して摂食している．このような摂食方法は他のグループの幼虫には見られない．Fは呼吸管がほとんど伸びてなく，常に水の表面に平行に静止し，時々頭部を180度回転して，口器を上にして口刷毛で水流を起こし，水の表面を浮遊する微細な有機物を口元に引き寄せて，咽頭内に取り込んでいる．大きな浮遊物は頭部を90度回転し，浮遊物を側面や下方に跳ね除けている．その様子は滑稽で見ていて飽きることがない．最後のGは呼吸管の先端が鋭くとがり，水生植物の根に呼吸管を刺し込んで根の組織内の酸素を吸収し呼吸している．幼虫A，D，Fと同じように口刷毛で水流を起こし，水中を浮遊する有機物を口元に寄せ付けて口腔内に取り込んでいる．水生植物の根に付着して幼虫・蛹の期間は水面に浮上しない．どのようなメリットがあるのだろうか．

上記のような生態観察によりグルーピングされた沖縄産の蚊の幼虫（AからG）を検索表で同定するとグループAはネッタイイエカ *Culex quinuqefasciatus*（動画1J），Bはヒトスジシマカ *Aedes albopictus*（動画2J），Cはトラフカクイカ *Lutzia vorax*（動画15J），Dはコガタアカイエカ *Cx. tritaeniorhynchus* またはニセシロハシイエカ *Cx. vishnui*，Eはカラツイエカ *Cx. bitaeniorhynchus*（動画6J），Fはシナハマダラカ *Anopheles sinensis*（動画3J），Gはアシマダラヌマカ *Mansonia uniformis*(動画9J)であった．幼虫が水面・水中・水底で採食している様子やカクイカが他の幼虫を捕食している様子は圧巻で見飽きることはない．我が国の蚊科の形態学的分類はよく研究されているが，肉眼による生態観察の結果だけでもある程度種の同定は可能である．

この文章は宮城（2023）が，ペストコントロール，204号，60頁（コラム）に若干追筆したものである．

Fun and usefulness of observing the feeding behavior of mosquito larvae

Mosquito larvae are commonly called "boufura" in Japanese. They have been called "stick-shaking" because they swim in the water by twisting their bodies like a stick. An empty discarded can had been left in the shady corner in my garden and was filled with rain water from the heavy rain a few days ago. Before I knew it, many mosquito larvae had appeared in it. I brought these mosquito larvae to my house along with the standing water and sediment from the bottom of the water, transferred them to several small aquariums (10 cm width × 7 cm hight × 2.5 cm depth), placed them next to my sofa where I usually rest, and began observing and photographing the mosquito larvae's movements while drinking tea. I also collected large numbers of mosquito larvae from nearby rice fields and wetland puddles. It became a daily routine for the past three years, just as people began staying at home due to the coronavirus pandemic.

All mosquito larvae except for the *Mansonia* larvae, breathe by protruding the spiracular opening of their breathing tube, siphon, above the water surface and taking in oxygen from the air. Therefore, the movement and feeding of mosquito larvae are closely related to their respiration modes. Most species of mosquito larvae are omnivorous, feeding on tiny organic matter of animals and plants floating on the surface or in the water, or on the carcasses of insects and other small animals found in the sediments on the bottom of the water.

When watching mosquito larvae swimming in an aquarium, I can see that they mainly swim in two ways. One is "side to side lashing movement", in which they wiggle their bodies and move between the surface and bottom of the water with their heads behind them. The other is "gliding movement", in which they move their heads without wiggling their bodies. The "side to side movement" is seen mainly when the larvae staying at the water surface are startled by vibrations, etc. In the "gliding movement", they move on the water surface by flapping their flexible mouth-brushes and taking in floating objects in the water into their oral cavity. In addition, the larval posture at the surface of the water when resting or breathing, swimming behavior, and feeding behavior vary slightly from individual to individual; they are characteristic of the genus or species.

All the 40 mosquito larvae collected in the empty can in the garden were in the 3rd or 4th stage. These larvae in the aquariums were distinguished by my naked eye into three groups, A, B, and C, based on their underwater swimming movement, static posture on the water surface, feeding method, and external morphology (size, color). Group A consisted of 22 individuals, with a head width of about 1 cm, long and slender respiratory tubes, and well-developed mouth-brushes. They remained motionless on the water surface, occasionally sliding sideways to vibrate their mouth-brushes and draw floating minute objects in the water to their mouths to feed by the collecting-filtering mode. It's adorable to see them hanging on the water surface and moving their mouth-brushes in unison.

Group B had 15 individuals, with the same head width as that of A, but the respiratory tube was thicker and apparently shorter than A. They wiggled up and down between the surface and bottom of the water, digging their heads into the sediments on the bottom and nibbling the organic matter. Sometimes they rose to the surface with small chunks of sediment in their mouths and consuming them by spinning the chunks like a windmill (Collecting-gathering mode). They looked just like a street performer in the square.

Group C consisted of only three larvae which were significantly larger than the larvae of the previous two groups. They had a Omusubi-shaped (triangle-shaped) head, with head-width of appromately 1 cm, and a thick, short siphon. They moved very little and bit and ate approaching prey larvae (Predation). Their predation reminded me of the lions I saw the other day in the African savanna preying on gazelles.

The larvae collected from rice fields and wetlands could be divided into four groups, D, E, F and G, which were different from the previous three groups collected in the container. The larvae of group D were very similar to group A in terms of head width, movement, and feeding method, but the siphon was clearly longer. Group E larvae also had a long siphon, similar in morphology to the larvae of groups A and D. The larvae were always feeding on green filaments in clumps of green algae, *Spirogyra*. This is why the intestines inside the larvae's abdomen are greenish in color. The larvae reeled in the long, thin threads of *Spirogyra* with their antennae and mouthparts, then split the thread into two pieces using their mandibles, maxilla, and dorsomentum. They kept one piece of the thread on their head and fed on the other piece by shredding it with their mandibles. This feeding method is not seen in the other groups of larvae.

In group F, the siphon was barely extended and the larva always remained stationary parallel to the water's surface. It occasionally rotated its head 180 degrees, placed its mouthparts upwards, and created a water current with its mouth-brushes to draw minute organic matter floating on the surface of the water to its mouth and into its pharynx. When confronted with a large floating matter, the larvae rotated their heads 90 degrees, throwing the floating matter to the side and downward. The behavior is comical and never boring to watch.

The siphon of the larvae of group G had a sharp tip that could be inserted into the roots of aquatic plants to absorb oxygen from the root tissue. Like the larvae of groups A, D, and F, they created a water current with their mouth-brushes, attracting organic matter floating in the water to their mouth and into their pharynx. They attached their respiratory organ (siphon) to the roots of aquatic plants and did not rise to the water surface during the larval and pupal periods. What advantage does this behavior have?

The mosquito larvae breeding in the Ryukyu Archipelago, grouped as A to G based on the ecological observations described above, were identified using the keys (Toma and Miyagi, 1986). Group A was *Culex quinquefasciatus* (Video 1E), B was *Aedes albopictus* (Video 2E), C was *Lutzia vorax* (Video 15E), D was *Cx. tritaeniorhynchus* or *Cx. vishnui,* E was *Cx. bitaeniorhynchus* (Video 6E), F was *Anopheles sinensis* (Video 3E), and G was *Mansonia uniformis* (Video 9E). The observation of the larvae feeding on the surface, underwater or on the bottom of the water, and the *Lutzia* larvae preying on other mosquito larvae, is a spectacular sight that is never boring. The morphological classification and biology of mosquitoes in the Ryukyu archipelago has been well studied, and it is possible to identify species to a certain extent based on the results of biological observations with the naked eye alone.

This article has updated slightly that of Miyagi (2023), Pest Control, No. 204, page 60 (Column).
(I. Miyagi)

水田　Paddy field

人工容器　Artificial containers

採集，飼育，標本作製の道具　Tools for collecting, rearing and preparing specimens

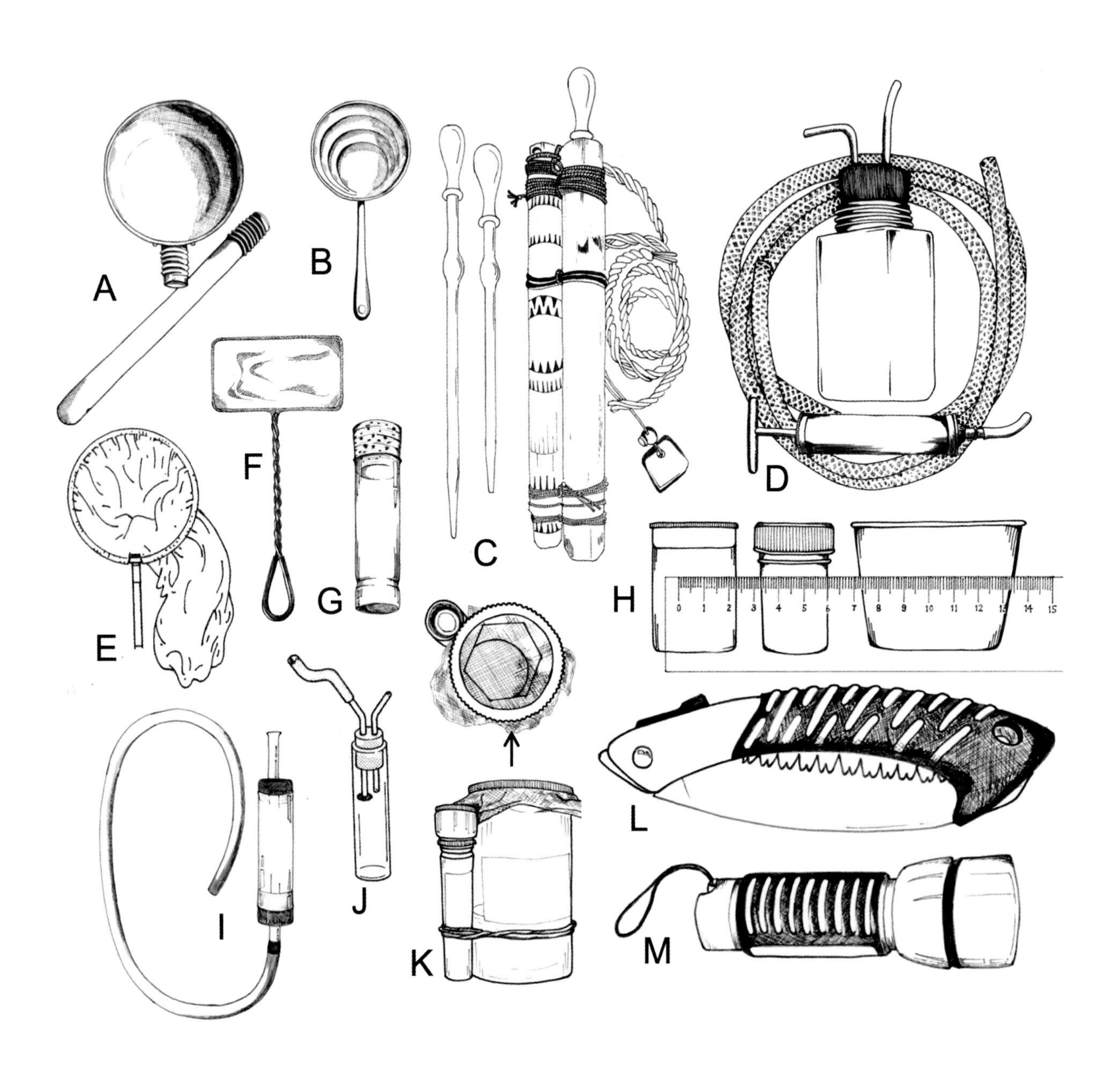

A．柄杓（並）Dipper；B．柄杓（小）Handy dipper；C．スポイドとスポイドケース（竹筒）Pipettes with case；D．携帯ポンプ一式 Handy suction pump；E．捕虫網 Insect sweeping net；F．金魚掬い網 fish scooping；G．毒瓶 Insect kiling bottle；H．各種プラスチック容器 Plastic bottles；I，J．吸虫管 Suction-bottle；K．個別飼育容器 Small bottle for individual rearing；L．折りたたみ式のこぎり Hand saw；M．懐中電器 Flashlight

第3章
初心者のための日本産蚊科幼虫の検索図

（主として4齢幼虫）

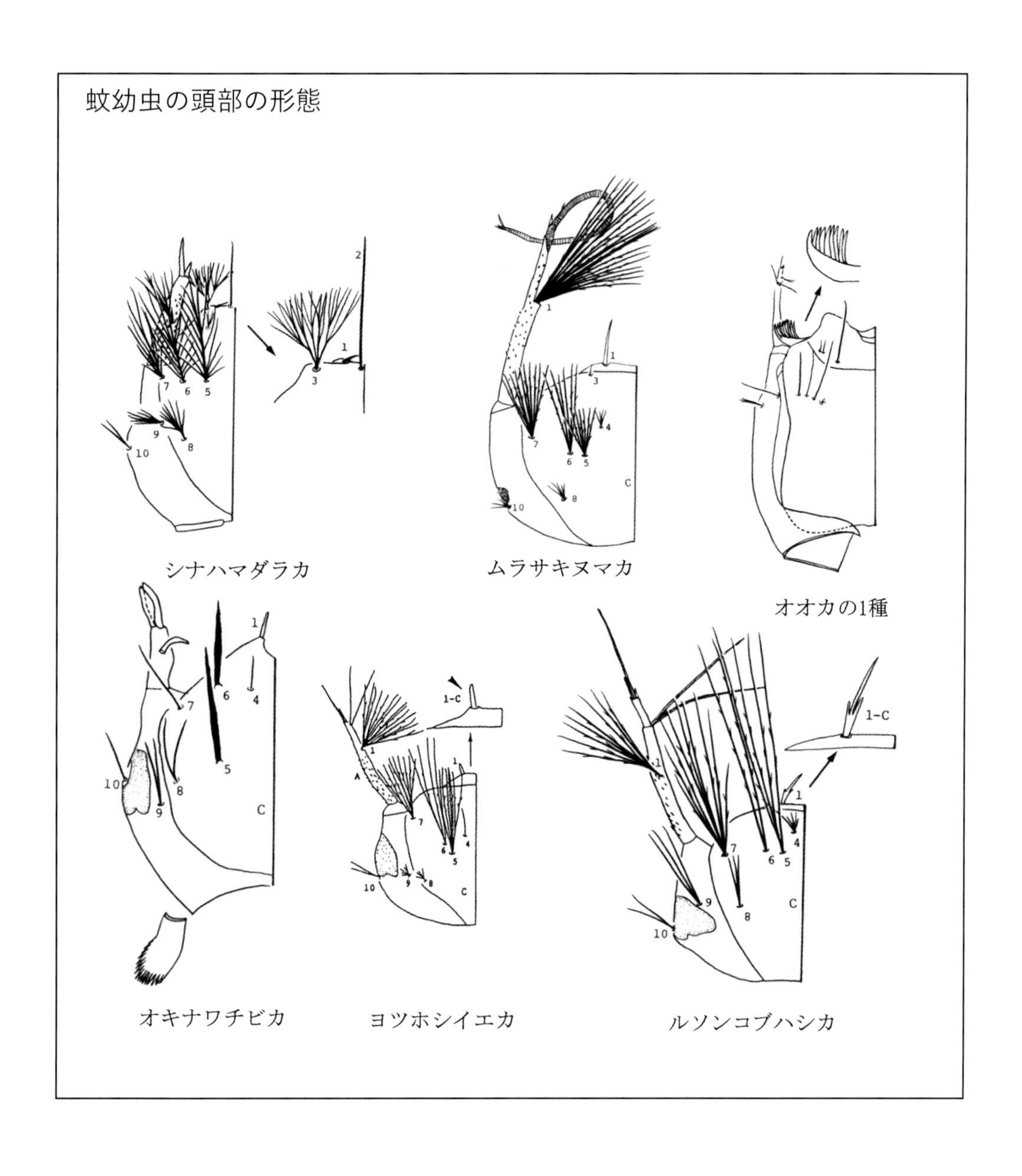

日本産蚊科幼虫の検索図を作成するに当たって　（水田英生）

私（水田）は検疫所内で蚊にかかわる研修を 10 年以上行っていたが，初心者には顕微鏡で見る標本像とイラストがなかなか一致せず，その都度図解または写真を撮り説明しなければならなかった．そこで所内の内部資料として「検疫所衛生技官のための日本に棲息する蚊の同定，成虫と幼虫編」を作成し活用してきた．津田（2019）はこの内部資料を基に各種蚊成虫の全身図や特徴を独特の手法で描写したオリジナルな図解検索表を作成された．いずれ幼虫編も作成する予定と聞いていたが，残念ながら道半ばにして逝去された．その後，私は機会あるごとに種の特徴を顕微鏡で撮影し，自作の「日本産蚊科幼虫の検索図」を完成させた．わが国にはイラスト図入りの日本産蚊科幼虫の検索表がまだないため，この機会に本検索図を初心者向けに公にすることにした．

使用上の注意

本検索図は初心者でも使えるようにできる限り多くの標本写真を用い，標本が手元にないものについては線画で示した．写真では線画と異なり重要な特徴を明確に示すことが困難である．標本は立体的であるため，写真では同一視野のすべての特徴に焦点を当てることができない．また，本検索図を印刷して用いると写真が小さいので，特徴が微小な場合は確認できない．したがって，本検索図はパソコン，タブレットあるいはスマートフォン等に保存し，必要に応じ写真を拡大して使用してもらいたい．また，本検索図と既存の検索表　（宮城・當間，2017；Tanaka et al., 1979；津田良夫 . 2019）と併用して検索していただきたい．種の同定は検索表の形態的特徴だけでなく，それぞれの種の分布(付表 2)，幼虫が発生している水溜まり（付表 3)，遊泳や摂食行動（動画 23）の特徴，吸血嗜好性など生態的特徴（付表 4 ）も含めて行うことが重要である．

PDF

初心者のための

日本産蚊科幼虫の検索図

(主として4齢幼虫)

本検索図は手元の標本と比べ易くするために多くの標本写真を用いた．そのため同定ポイントが誇張できず，掲載写真では体毛等が分かりにくくなってしまった．今回，検索図をPDFファイルにし，分かりにくいところは倍率を上げて見らるようにし，極力同定ポイントを見やすくした．分かりにくいところは倍率を上げて利用していただきたい．

なお，初心者でも分かり易くするために，部位の名称は簡単なものにした．

2024年7月
神戸検疫所ベクター室
水田英生

検索表に出てくる名称と部位（ナミカ族）

胸部
前胸：P　前胸毛0：0-P・・・・・前胸毛14：14-P
中胸：M　中胸毛0：0-M・・・・・中胸毛14：14-M
後胸：T　後胸毛0：0-T・・・・・後胸毛13：13-T
腹部(第Ⅰ腹節～第Ⅵ腹節)
第Ⅰ腹節：Ⅰ　第Ⅰ腹節毛1：1-Ⅰ・・・・第Ⅰ腹節毛14：14-Ⅰ
第Ⅱ腹節：Ⅱ　第Ⅱ腹節毛0：0-Ⅱ・・・・第Ⅱ腹節毛14：14-Ⅱ
：
：
第Ⅵ腹節：Ⅵ　第Ⅵ腹節毛0：0-Ⅵ・・・第Ⅵ腹節毛14：14-Ⅱ

触角：A
触覚毛1：1-A・・・・・触覚毛6：6-A
頭部：C
頭毛1：1-C・・・・・・頭毛15：15-C
大腮　：Mn　　小腮　：Mx
下唇板：MP　　小腮軸節基部　：MPlp
末端部：(第Ⅶ腹節～第Ⅹ腹節)
第Ⅶ腹節：Ⅶ～第Ⅷ腹節
第Ⅶ腹節毛0：0-Ⅶ
・・・・第Ⅶ腹節毛14：14-Ⅶ
第Ⅷ腹節毛0：0-Ⅷ
・・・・第Ⅷ腹節毛14：14-Ⅷ

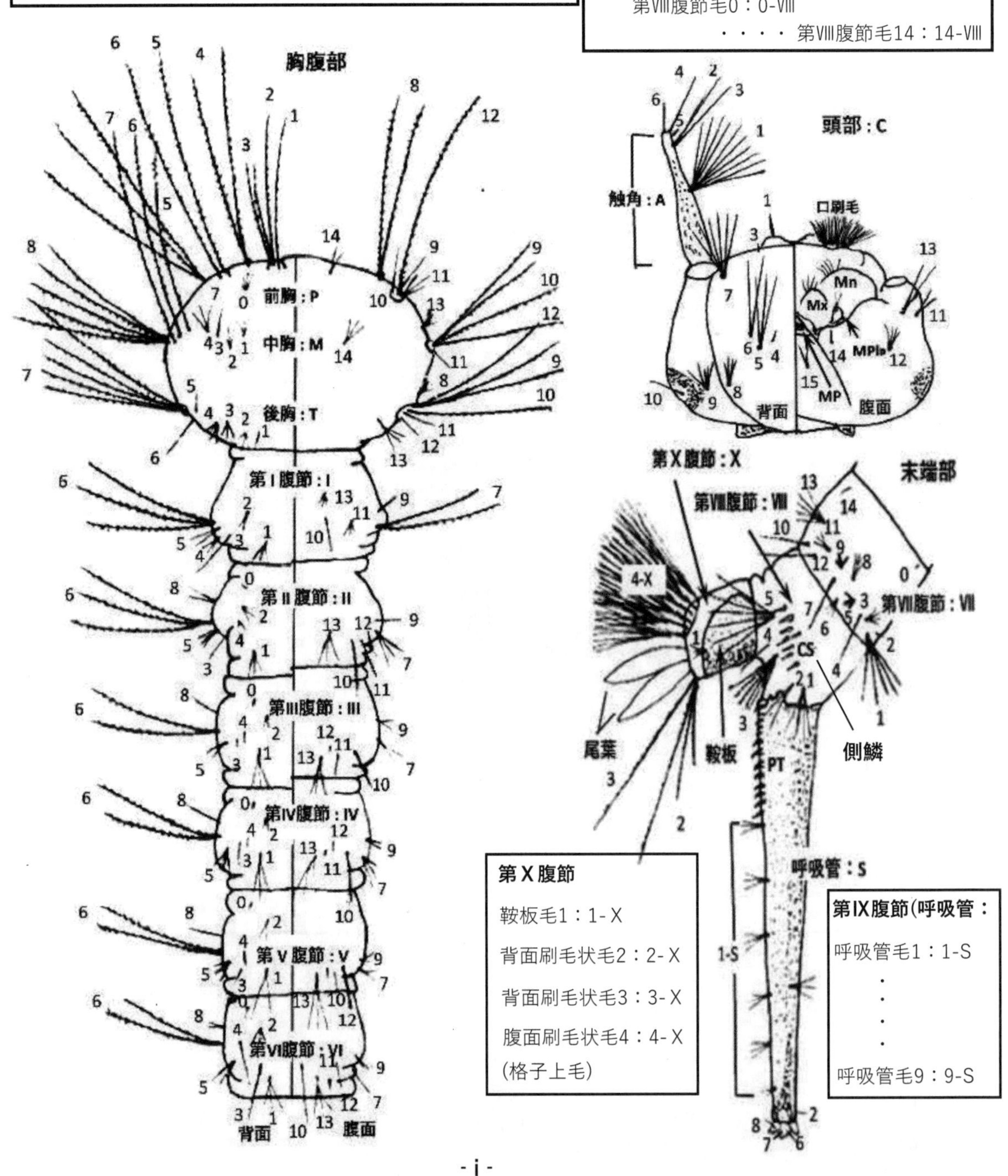

検索表に出てくる名称と部位（ハマダラカ族）

胸部
前胸：P　前胸毛0：0-P・・・・・前胸毛14：14-P
中胸：M　中胸毛0：0-M・・・・・中胸毛14：14-M
後胸：T　後胸毛0：0-T・・・・・後胸毛13：13-T
腹部(第Ⅰ腹節～第Ⅵ腹節)
第Ⅰ腹節：I　第Ⅰ腹節毛1：1-I
・・・・第Ⅰ腹節毛14：14-I
第Ⅱ腹節：II　第Ⅱ腹節毛0：0-II
・・・・第Ⅱ腹節毛14：14-II
：
：
第Ⅵ腹節：VI　第Ⅵ腹節毛0：0-VI
・・・・第Ⅵ腹節毛14：14-VI

(第Ⅲ腹節毛1～第Ⅷ腹節毛1：掌状毛)

触角：A
触覚毛1：1-A・・・・・触覚毛6：6-A
頭部：C
頭毛0：0-C・・・・・・頭毛15：15-C
小腮毛1　：1-Mx
下唇板：MP
末端部：(第Ⅶ腹節～第Ⅹ腹節)
第Ⅶ腹節：Ⅶ～第Ⅷ腹節
第Ⅶ腹節毛0：0-Ⅶ・・・・第Ⅶ腹節毛14：14-Ⅶ
第Ⅷ腹節毛1：1-Ⅷ・・・・第Ⅷ腹節毛5：5-Ⅷ
第Ⅹ腹節：X
鞍板毛：1-X
背面刷毛状毛：2-X
背面刷毛状毛：3-X
腹面刷毛状毛(格子上毛)4：4-X

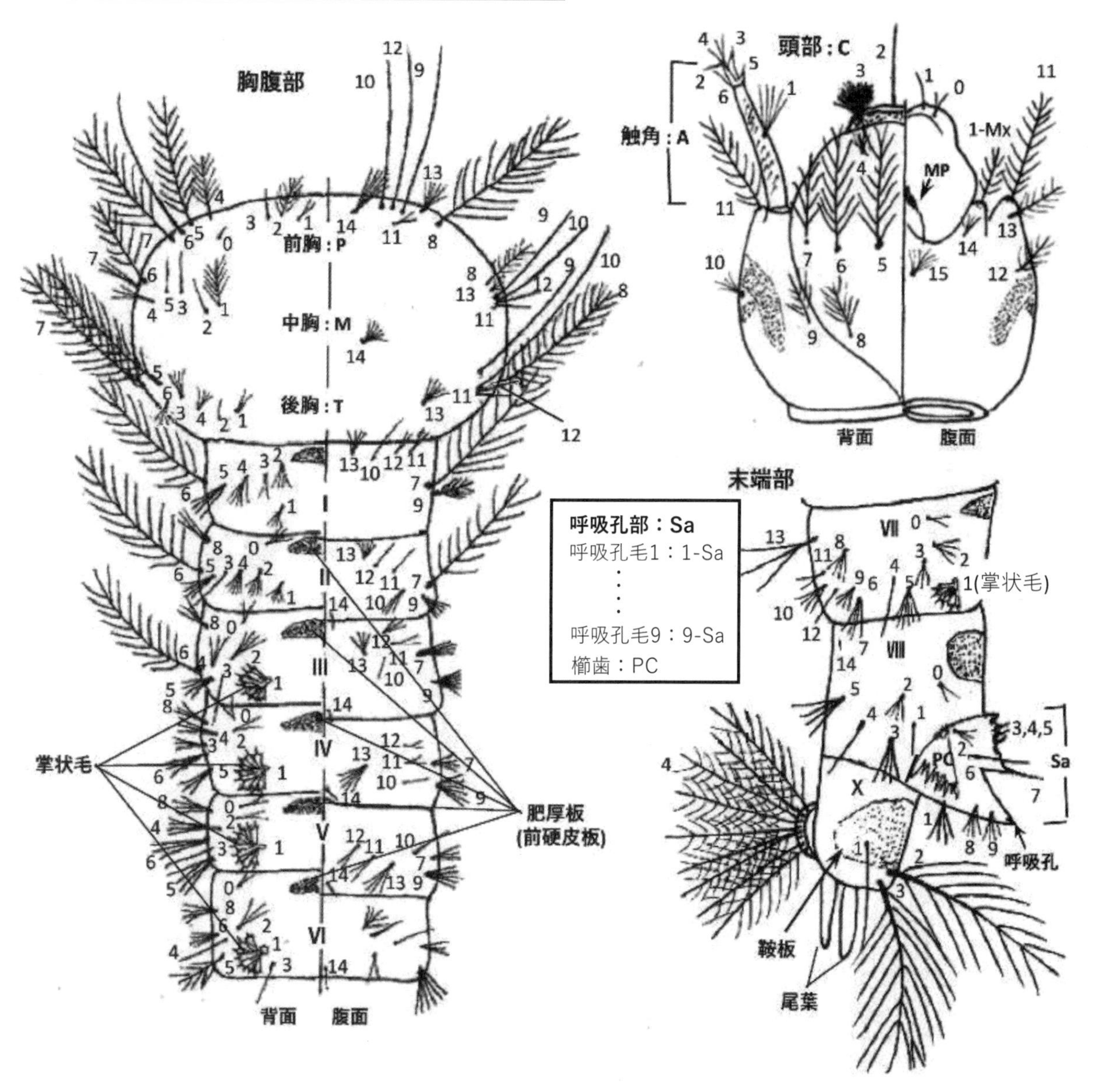

日本産蚊科幼虫（第４齢虫）

呼吸管を欠く．腹節毛１(1-I~VII)の多くは掌状毛．

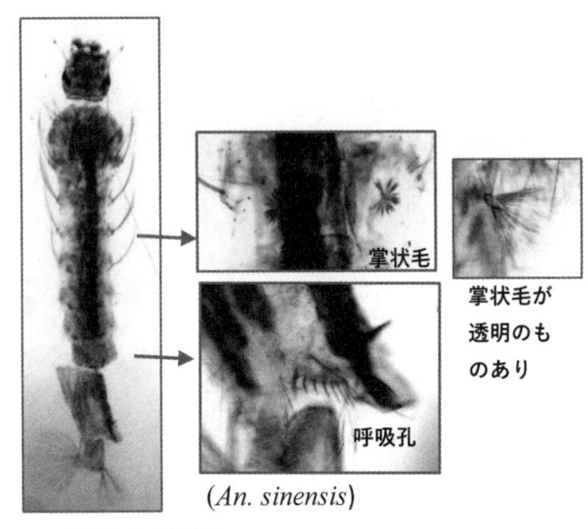

(*An. sinensis*)

ハマダラカ亜科 ANOPHELINAE
（５ページAへ）

呼吸管はよく発達する．腹節毛１(1-I~VII)の多くは掌状を呈さない．

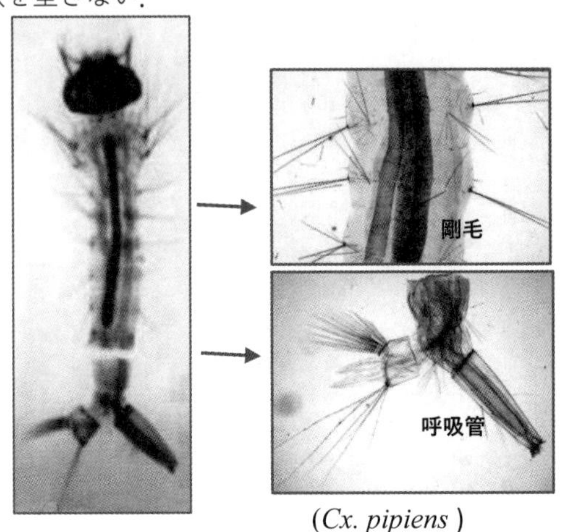

(*Cx. pipiens*)

ナミカ亜科 CULICINAE

極めて大型．口刷毛は6~約10本の扁平な鎌状片に変形する．側鱗(CS)及び呼吸管棘(PT)を欠く．

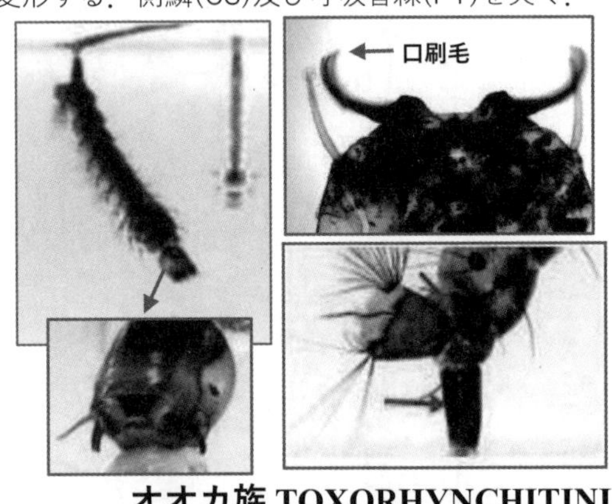

オオカ族 TOXORHYNCHITINI

オオカ属 *Toxorhynchites*

中型ないし小型．口刷毛は多数の繊細毛（カクイカ属は太い口刷毛）．側鱗(CS)及び呼吸管棘(PT)を有するものが大半．

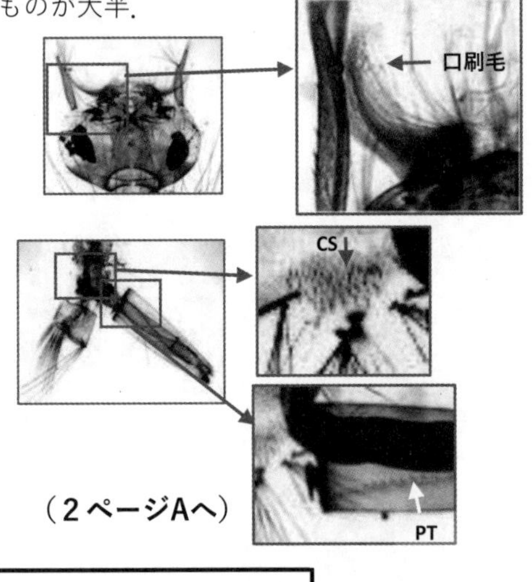

（２ページAへ）

中胸毛3,4（3,4-M）は背側の大きな肥厚板から別れた小さな肥厚板上にある．

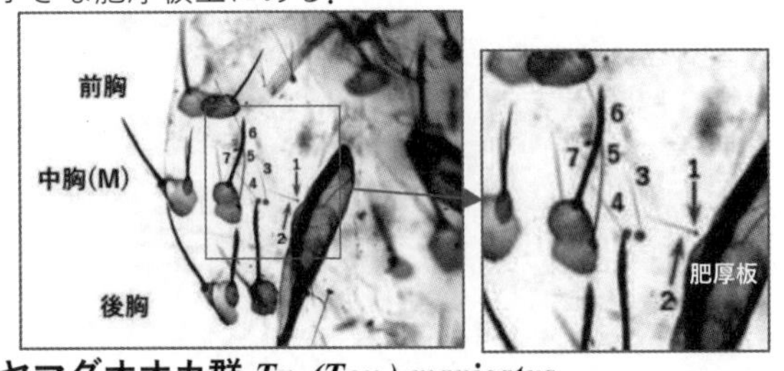

ヤマダオオカ群 *Tx. (Tox.) manicatus*
ヤマダオオカ *Tx. manicatus yamadai*（奄美群島に生息）
ヤエヤマオオカ *Tx. man. yaeyamae*（八重山群島に生息）

中胸毛3,4（3,4-M）は背側の大きな肥厚板にある．

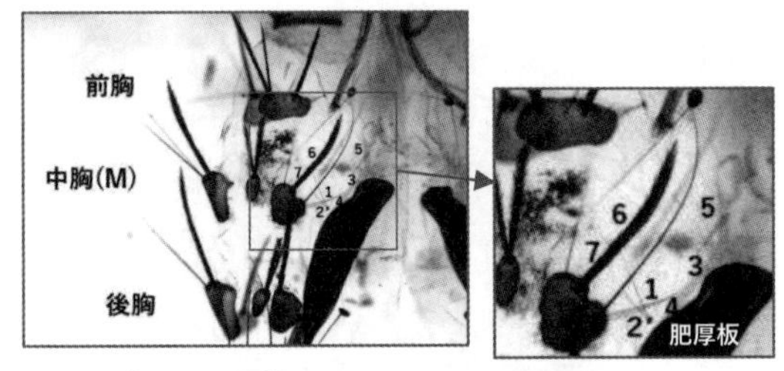

トワダオオカ群 *Tx. (Tox.) towadensis*
トワダオオカ *Tx. towadensis*（琉球列島を除く地域）
オキナワオオカ *Tx. okinawensis*（沖縄本島に生息）

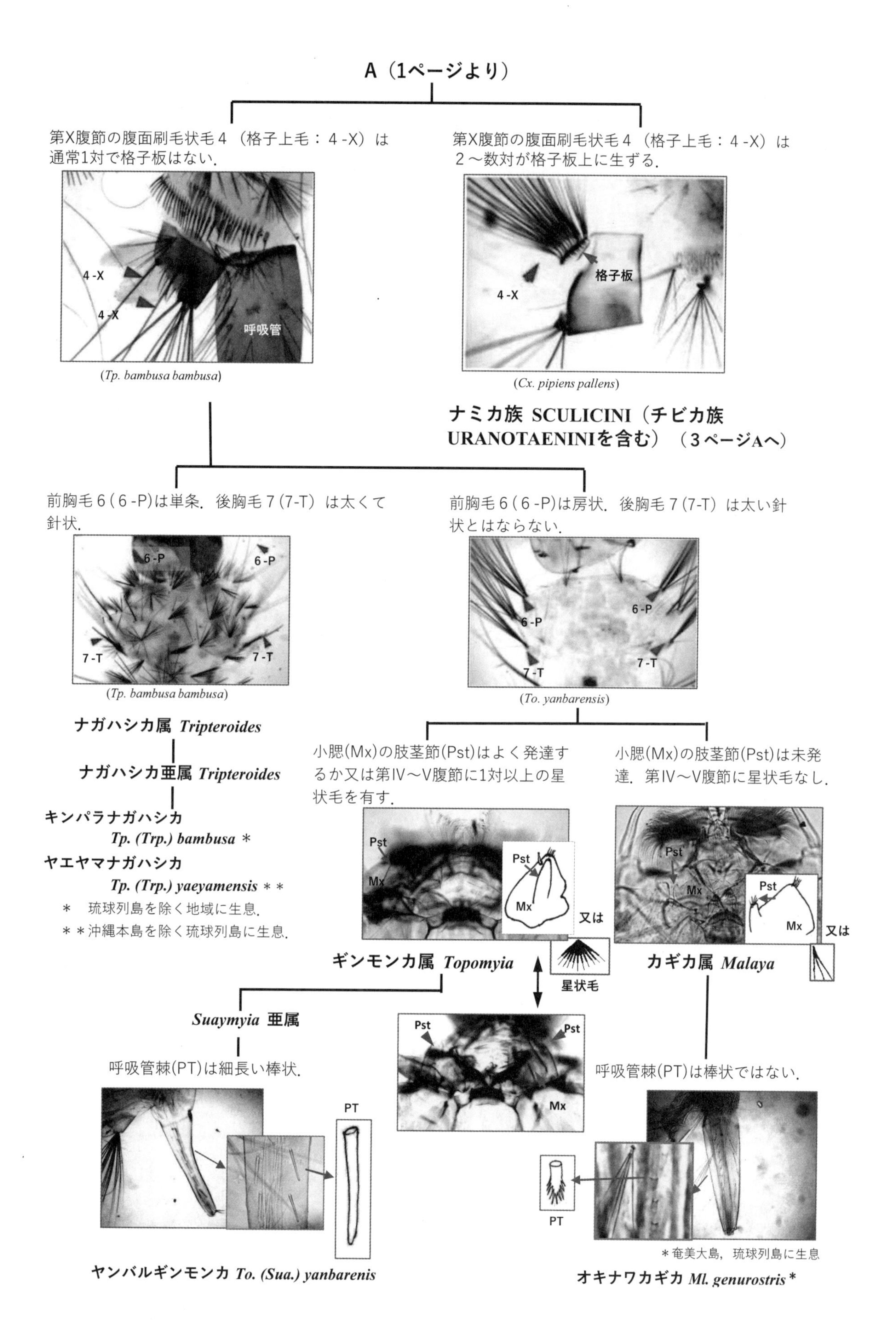
A（1ページより）
第X腹節の腹面刷毛状毛４（格子上毛：４-X）は
通常1対で格子板はない.
4-X
4-X
呼吸管
(Tp. bambusa bambusa)
第X腹節の腹面刷毛状毛４（格子上毛：４-X）は
２～数対が格子板上に生ずる.
格子板
4-X
(Cx. pipiens pallens)
ナミカ族 SCULICINI（チビカ族
URANOTAENINIを含む）（3ページAへ）
前胸毛６(６-P)は単条．後胸毛７(7-T）は太くて
針状.
6-P
6-P
7-T
7-T
(Tp. bambusa bambusa)
前胸毛６(６-P)は房状．後胸毛７(7-T）は太い針
状とはならない.
6-P
6-P
7-T
7-T
(To. yanbarensis)
ナガハシカ属 Tripteroides
ナガハシカ亜属 Tripteroides
キンパラナガハシカ
Tp. (Trp.) bambusa ＊
ヤエヤマナガハシカ
Tp. (Trp.) yaeyamensis ＊＊
＊　琉球列島を除く地域に生息.
＊＊沖縄本島を除く琉球列島に生息.
小腮(Mx)の肢茎節(Pst)はよく発達す
るか又は第IV～V腹節に1対以上の星
状毛を有す.
Pst
Mx
Pst
Mx
又は
星状毛
ギンモンカ属 Topomyia
小腮(Mx)の肢茎節(Pst)は未発
達．第IV～V腹節に星状毛なし.
Pst
Mx
Pst
Mx
又は
カギカ属 Malaya
Suaymyia 亜属
呼吸管棘(PT)は細長い棒状.
PT
Pst
Pst
Mx
呼吸管棘(PT)は棒状ではない.
PT
＊奄美大島，琉球列島に生息
ヤンバルギンモンカ To. (Sua.) yanbarenis
オキナワカギカ Ml. genurostris＊
-2-

A（2ページより）

ナミカ族 CULICINI（チビカ族 URANOTAENIINIを含む）

呼吸管は先端が尖った円錐形で鋸歯を持ち植物の根に突き刺して呼吸する．

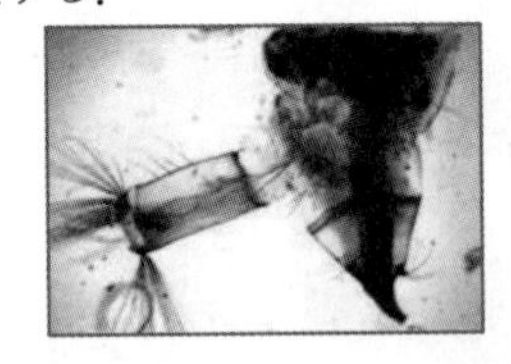

ヌマカ属 *Mansonia* 及びキンイロヌマカ属 *Coquillettidia*

呼吸管はほぼ平らな円筒状で水面で呼吸する

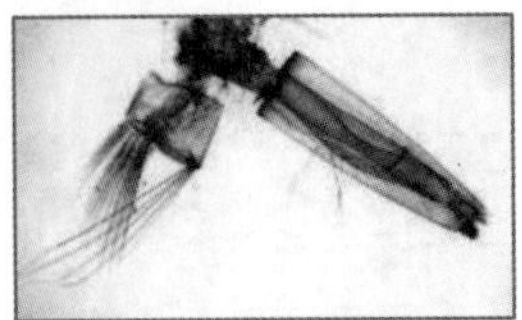

(*Cx. pipiens*)

触角の先端側半分は鞭状に長く伸び，少なくとも頭幅の1.5倍以上．側鱗（CS)は5～10個で先端は尖る．

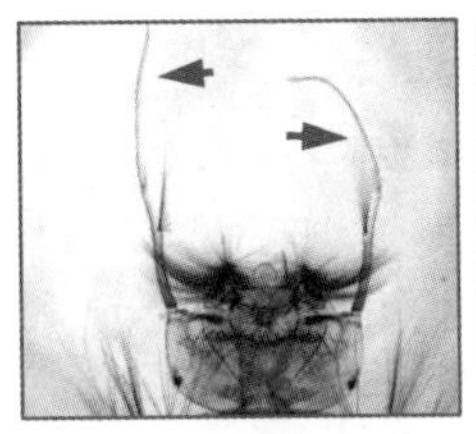

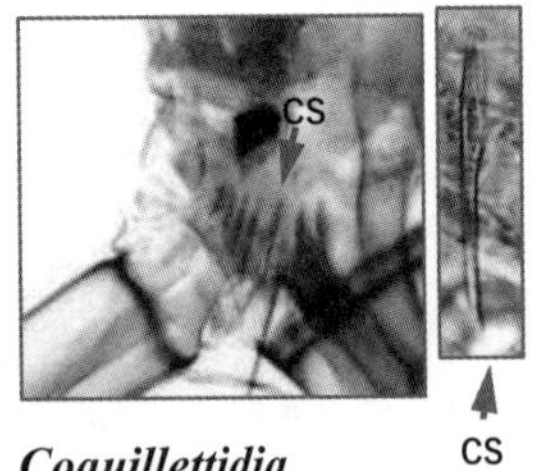

キンイロヌマカ亜属 *Coquillettidia*

触角の先端は鞭状に伸びず，頭幅とほぼ同長．側鱗（CS)は１～３個で先端は丸い．

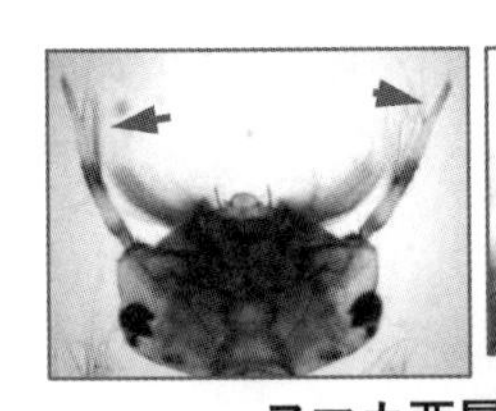

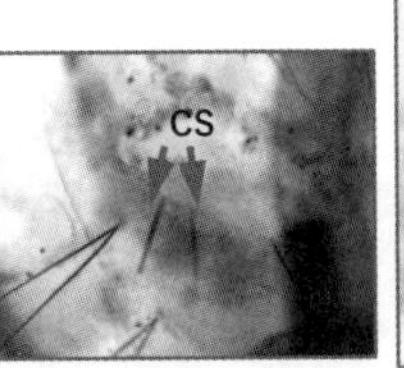

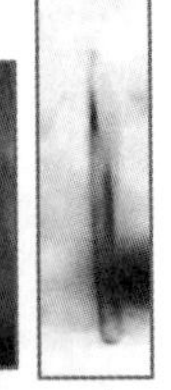

ヌマカ亜属 *Mansonioides*

アシマダラヌマカ *Ma.* (*Mnd.*) *uniformis*

側鱗(CS)は小さく亀の手状．鞍板には鞍板毛１(1-X)の他に1～2本の微剛毛を伴うこともある．

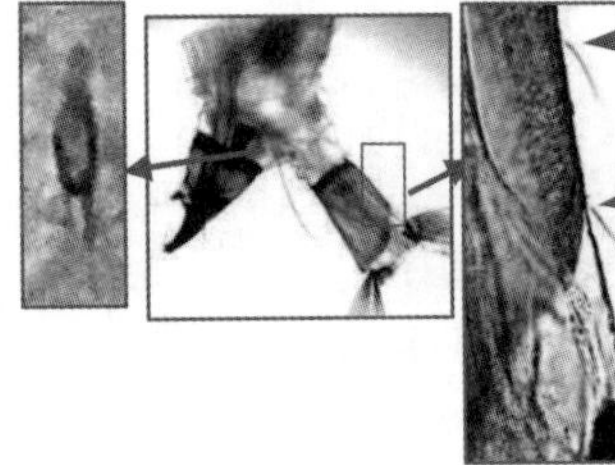

キンイロヌマカ
Cq.* (*Coq.*) *ochracea

側鱗(CS)は大きく細長い針状．鞍板には鞍板毛１(1-X)の他に3対の分岐毛を伴う．

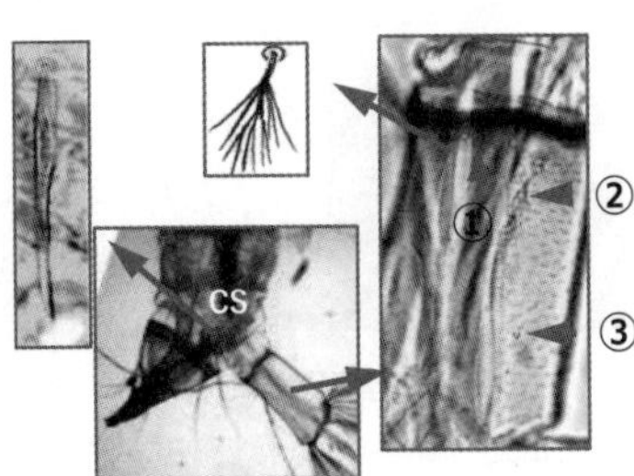

ムラサキヌマカ
Cq.* (*Coq.*) *crassipes

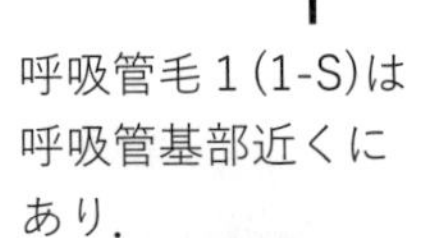

呼吸管毛１(1-S)は呼吸管基部近くにあり．

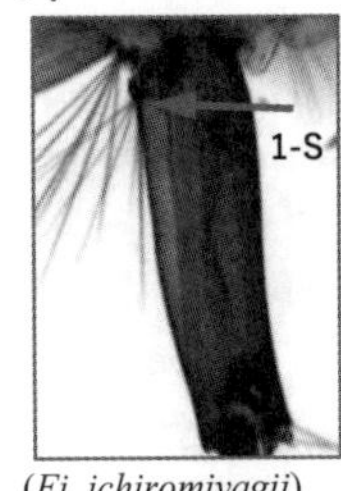

(*Fi. ichiromiyagii*)

頭毛１(1-C)は太く先端は尖る．

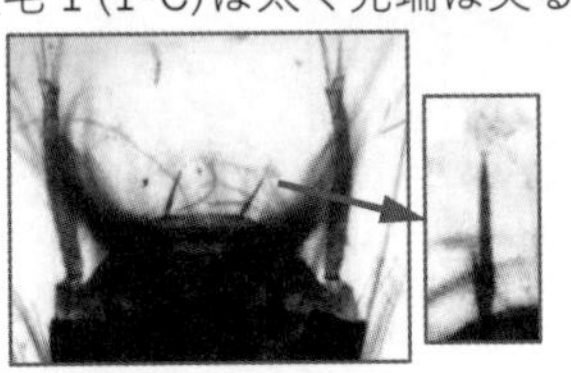

エセコブハシカ属 *Ficalbia*

オキナワエセコブハシカ
Fi. ichiromiyagii

頭毛１(1-C)は細く先端は糸状．

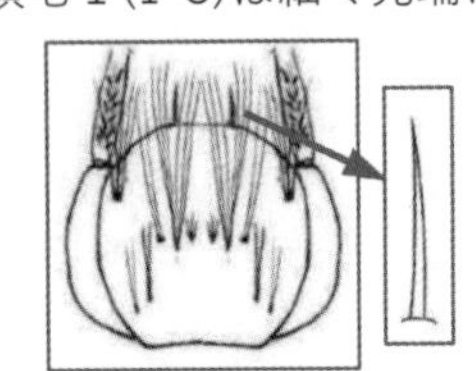

ハボシカ属 *Culiseta*
（４ページAへ）

呼吸管毛１(1-S)は呼吸管基部近くになし．

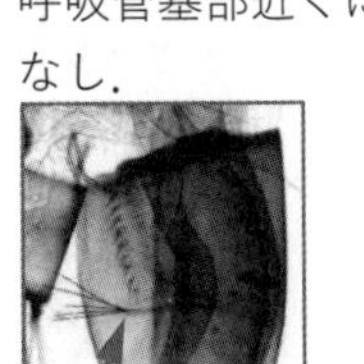

(*Ae. riversi*)

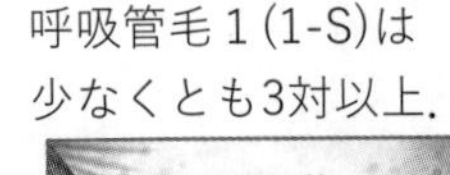

呼吸管毛１(1-S)は少なくとも3対以上．

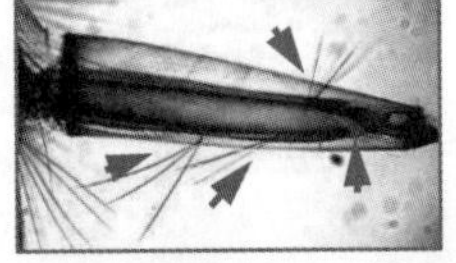

イエカ属 *Culex* 及びカクイカ属 *Lutzia*
（7ページAへ）

呼吸管毛１(1-S)は1対．

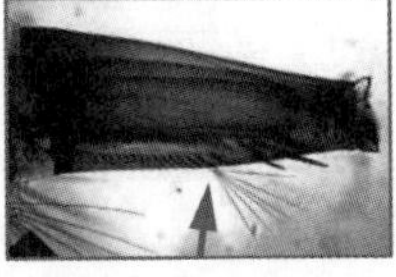

（４ページBへ）

A（3ページより）

ハボシカ属 *Culiseta*

触角は頭幅とほぼ同長. 触角毛１(1-A)は先端側1/3に生える. 呼吸管中央に毛列なし.

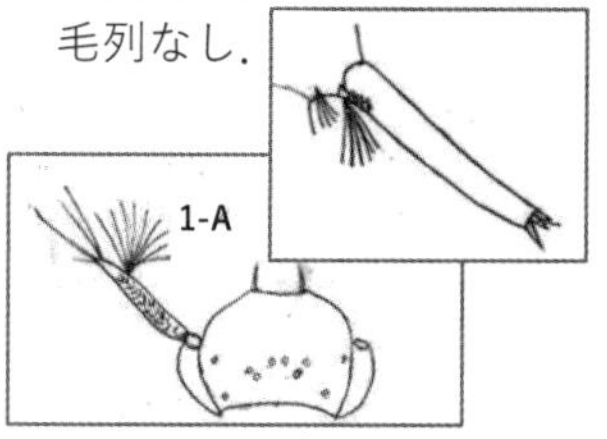

ヤマトハボシカ
Cs. (Cus.) nipponic

触角は頭幅の約1/2長. 触角毛１(1-A)は中央より基部側に生える. 呼吸管中央に毛列あり.

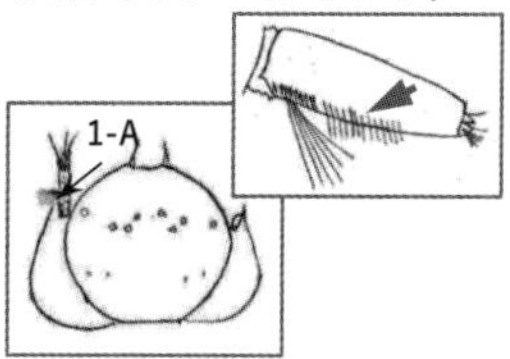

ミズジハボシカ
Cs. (Cus.) kanayamensis

（ハボシカ 亜属 *Culiseta*）

B（3ページより）

触角は触角毛2,3(2-,3-A)の直後にくびれて偽関節となる. 呼吸管棘(PT)を欠く.

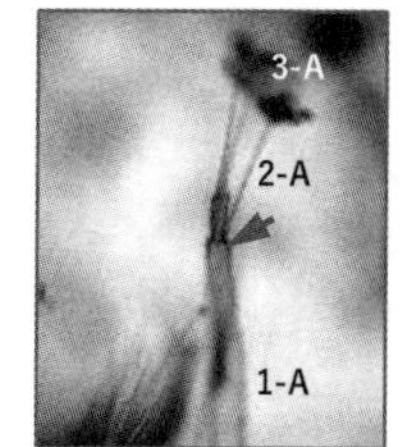

コブハシカ属 *Mimomyia*
Etorleptiomyia 亜属

側鱗(CS)は１列. 頭毛1(1-C)は３叉型.

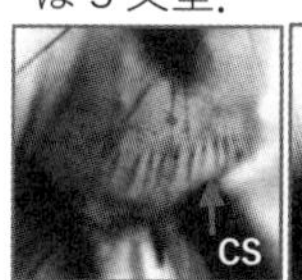

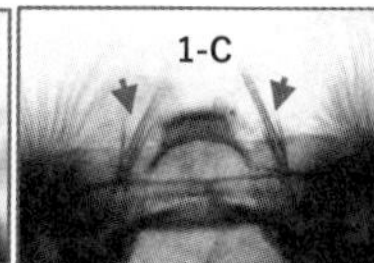

ルソンコブハシカ
Mi. (Eto.) luzonensis

側鱗(CS)は２列. 頭毛1(1-C)は単条.

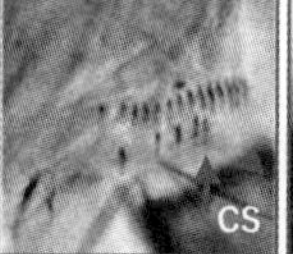

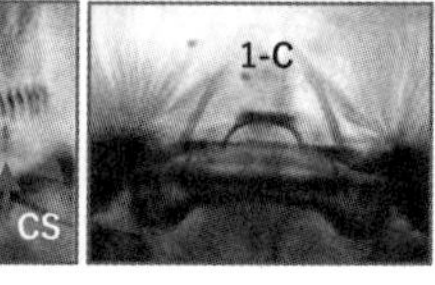

マダラコブハシカ
Mi. (Eto.) elegans

触角の先端部に偽関節構造なし. 呼吸管棘(PT)を欠くものもある.

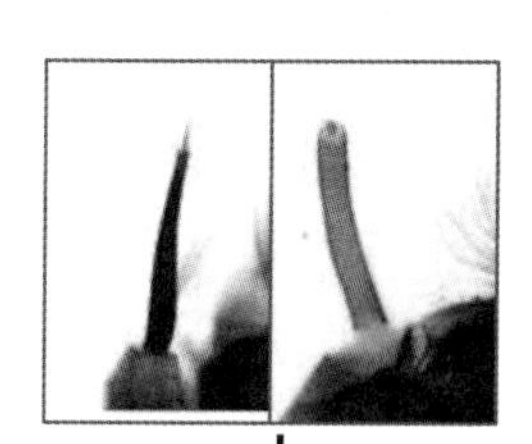

呼吸管棘(PT)を欠く

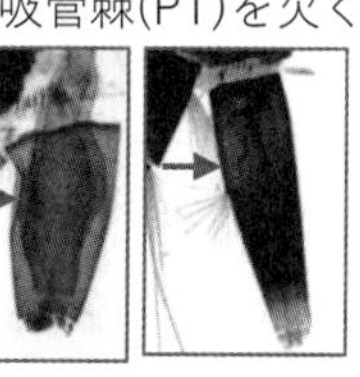

第Ⅷ腹節の肥厚板に大小同型の側鱗(CS)が2列に並ぶ.

ナガスネカ属
Orthopodomyia

第Ⅶ, 第Ⅷ腹節は暗褐色の肥厚板で覆われる. 鞍板毛1(1-X)は2〜3分岐. 背面刷毛状毛2(2-X)の1本は他に比べ非常に長い.

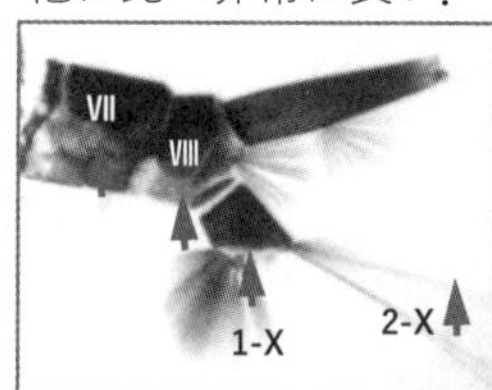

ハマダラナガスネカ
Or. anopheloides

第Ⅷ腹節に肥厚板はなく, 形状が同じの側鱗(CS)が1列ぶ.

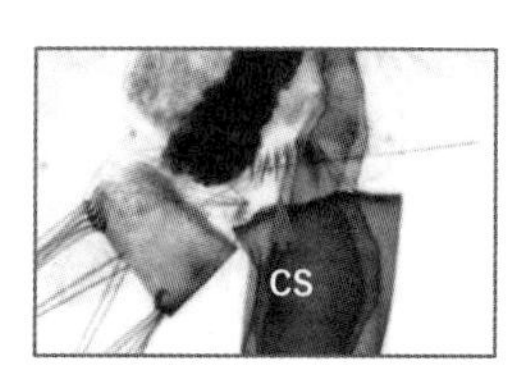

クロヤブカ属
Armigeres

側鱗(CS)は穂先型であるが先端は丸く鋸歯を伴う.

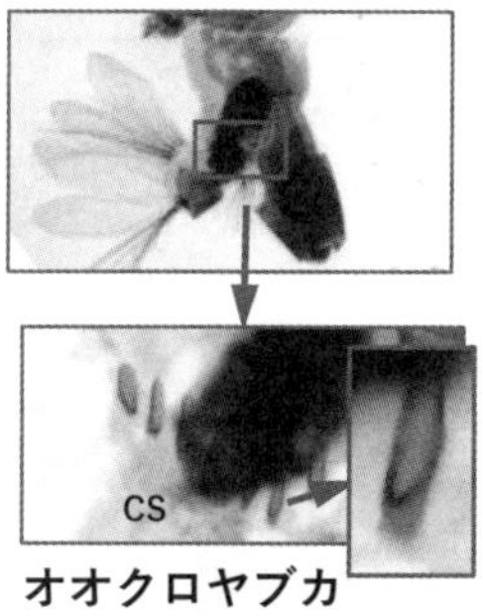

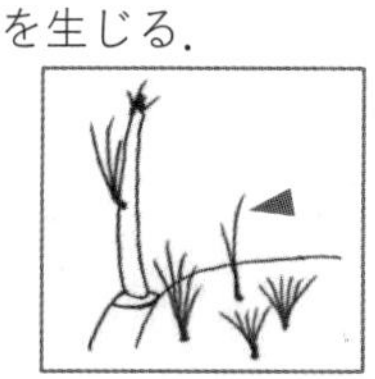

オオクロヤブカ
Ar. (Arm.) subalbatus
（クロヤブカ 亜属 *Armigeres*）

呼吸管棘(PT)を有す.

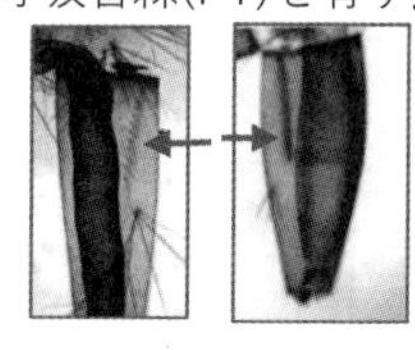

第Ⅷ腹節に肥厚板あり. 呼吸管棘(PT)の多くは鱗状.

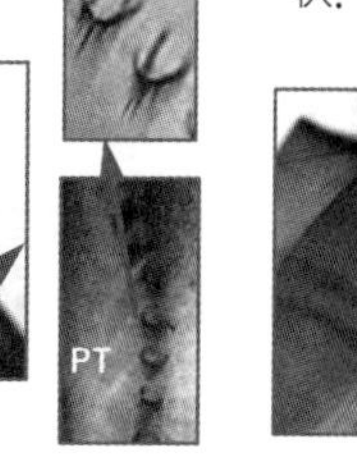

チビカ族
URANOTAENINI
チビカ属 *Uranotaenia*
（25ページBへ）

第Ⅷ腹節に肥厚板なし. 呼吸管棘(PT)は歯状または針状.

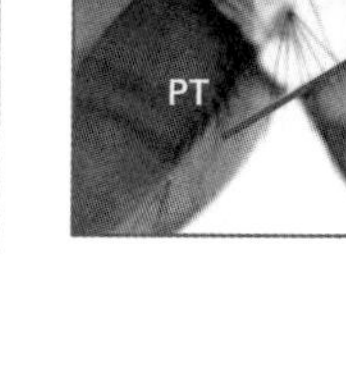

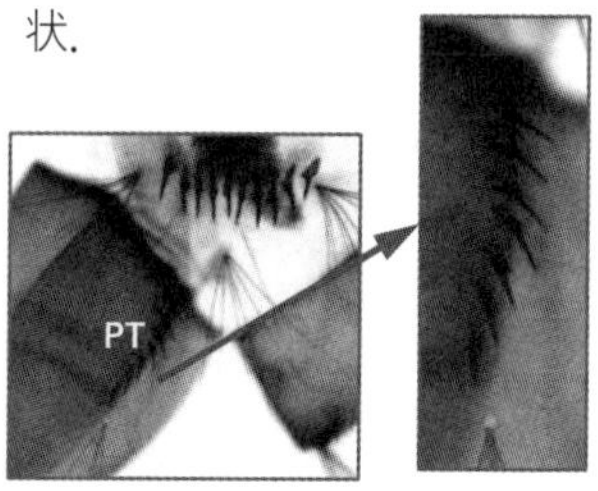

頭毛6(6-C)は単条で主幹に1本から数本の小側枝を生じる.

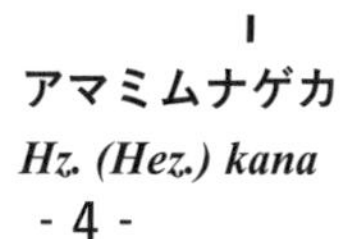

ムナゲカ属 *Heizmannia*
ムナゲカ 亜属 *Heizmannia*

アマミムナゲカ
Hz. (Hez.) kana

頭毛6(6-C)は単条若しくは2分岐か頭毛4(4-C)の様にほぼ等しく分岐するか扇状に分岐する.

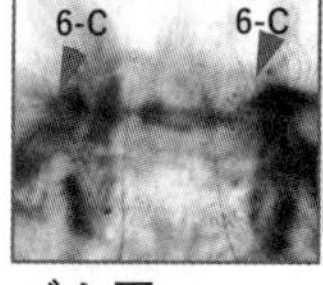

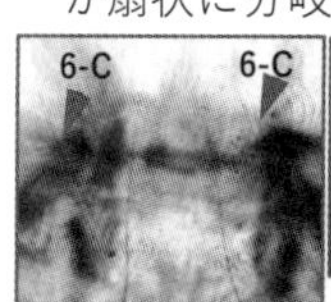

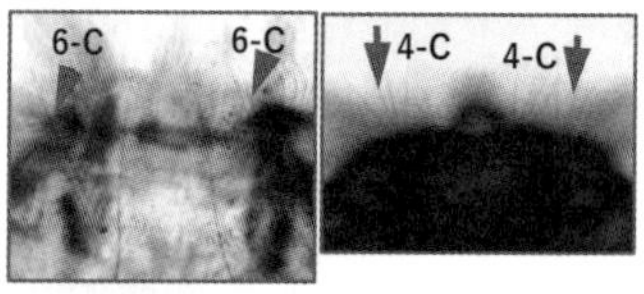

ヤブカ属 *Aedes*
フトオヤブカ属 *Verrallina*（広義
（14ページAへ）

- 4 -

A（1ページより）
ハマダラカ亜科 ANOPHELINAE
ハマダラカ属 *Anopheles*

頭毛2(2-C)の左右の間隔は頭毛2(2-C)と頭毛3(3-C)との間隔より広い.

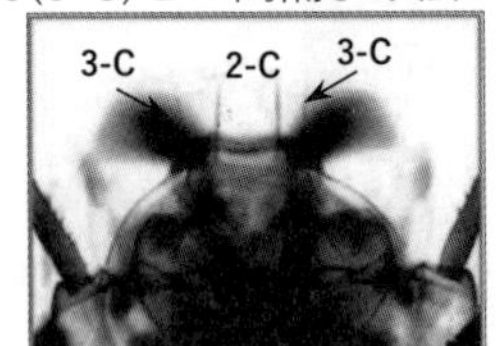

タテンハマダラカ亜属 *Cellia*

頭毛2(2-C)の左右の間隔は頭毛2(2-C)と頭毛3(3-C)との間隔より同等か狭い.

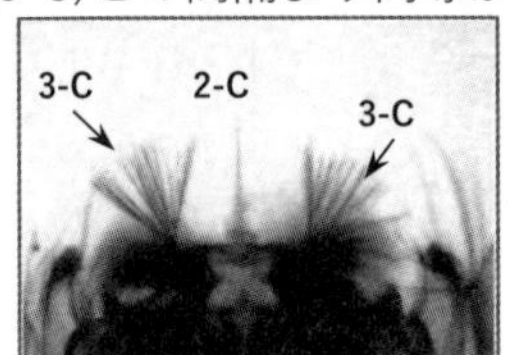

ハマダラカ亜属 *Anopheles*

腹部背面の肥厚板（前硬皮板）は左右の掌状毛からの間隔とほぼ胴長かそれより広い. 頭毛2(2-C)に小棘はない.

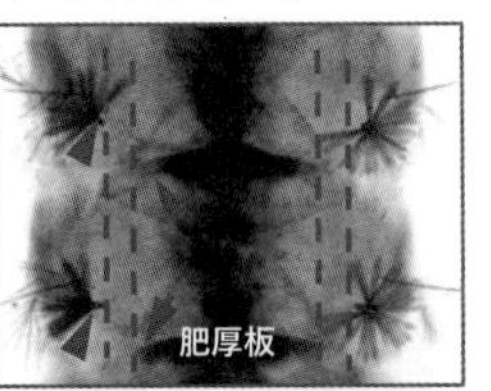

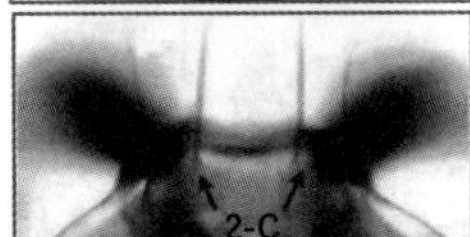

ヤエヤマコガタハマダラカ
An. (Cel.) yaeyamaensis

腹部背面の肥厚板（前硬皮板）は左右の掌状毛からの間隔より狭い. 頭毛2(2-C)に小棘がある.

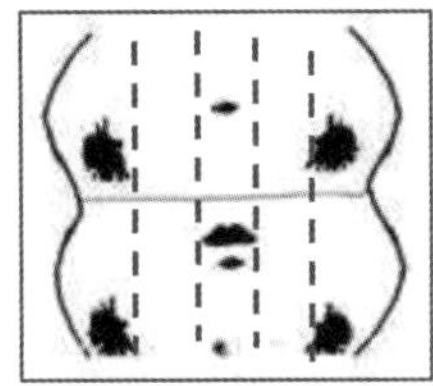

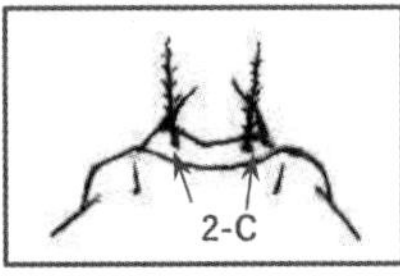

タテンハマダラカ
An. (Cel.) tessellatus

頭毛2(2-C)は4～7分岐し頭毛2(2-C)と頭毛3(3-C)の左右の間隔はほぼ等しい.

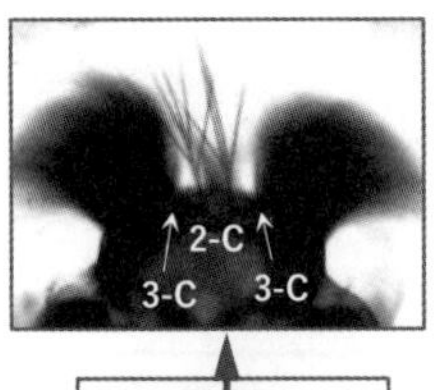

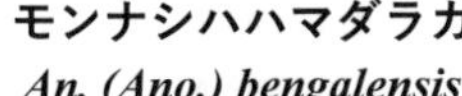

モンナシハハマダラカ
An. (Ano.) bengalensis

頭毛2(2-C)は単条で, 左右は接近する.

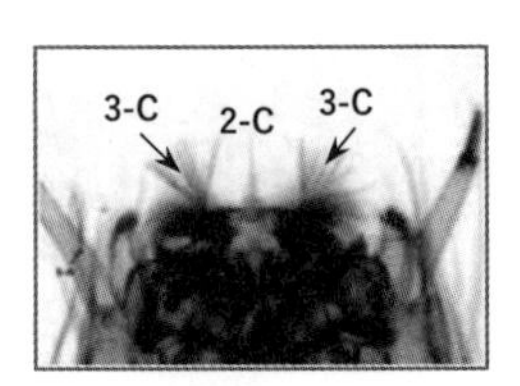

触角は微棘を伴わない. 触角毛1(1-A)と頭毛5-7(5-7-C)は単条.

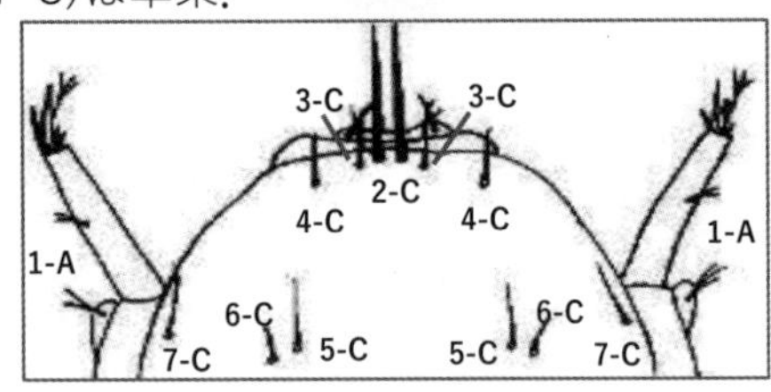

オオモリハマダラカ
An. (Ano.) omorii

触角は微棘を伴う. 触角毛1(1-A)と頭毛5-7(5-7-C)は分岐し, 頭毛5-7(5-7-C)は羽状.

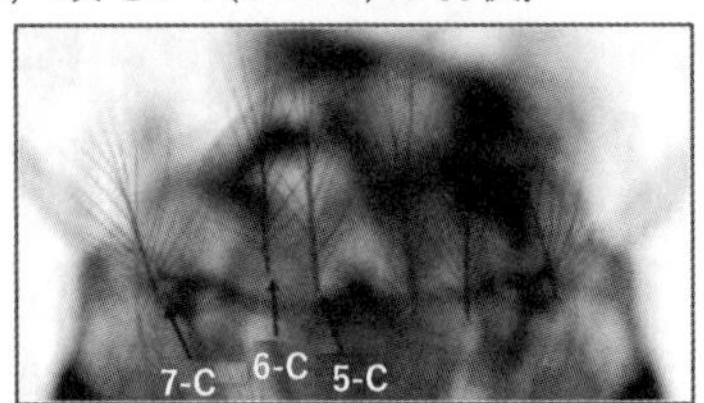

頭毛3(3-C)は単条. 頭毛8(8-C)は単条又は2分岐. 第Ⅰ腹節毛1(1-Ⅰ)は1～8分岐（通常1～4分岐）の単純な剛毛.

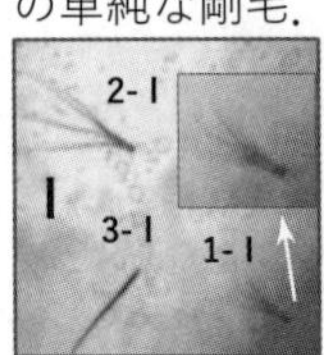

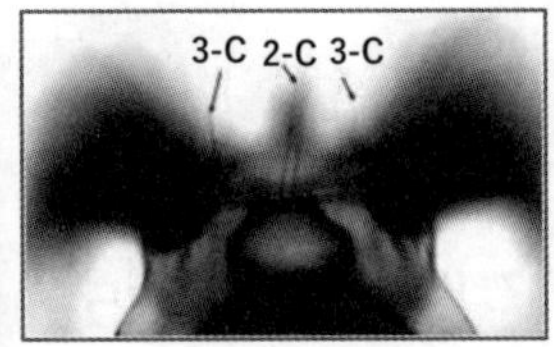

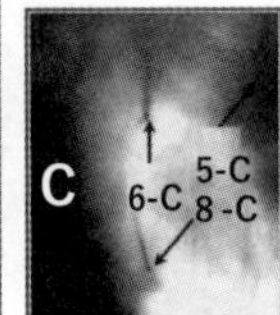

ヤマトハマダラカ ***An. (Ano.) lindesayi japonicus***

頭毛3(3-C)は少なくとも4分岐. 頭毛8(8-C)は少なくとも5分岐. 第Ⅰ腹節毛1(1-Ⅰ)は多くの細葉状の分枝を持つ.

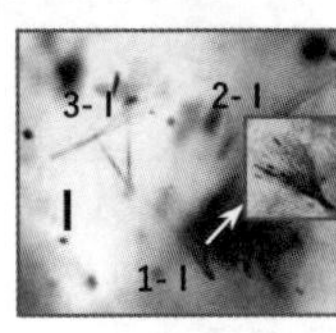

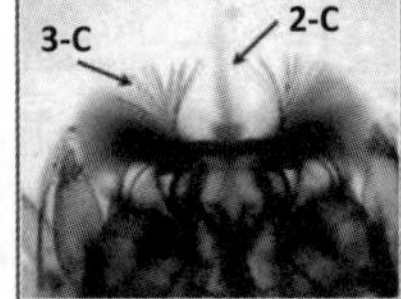

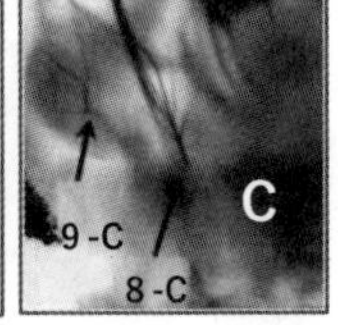

（6ページAへ）

A（5ページより）

頭毛3(3-C)は8以下に分岐する。第Ⅱ～Ⅶ腹節毛０（0-Ⅱ～0-Ⅶ）は単条かつ微小.

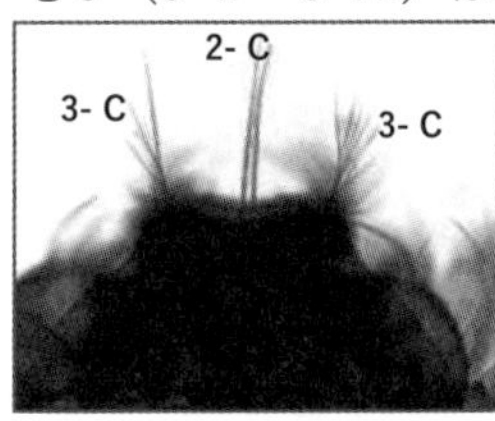

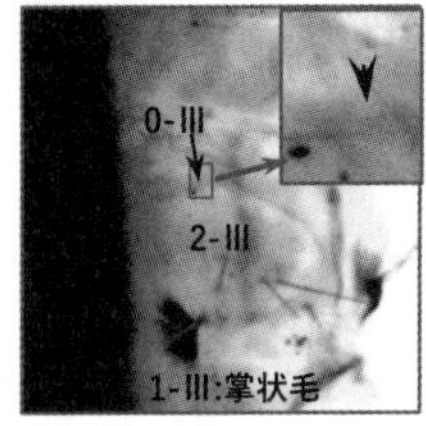

（蛹での同定が正確ではあるが，下記により推定は可能）

頭毛3(3-C)は頭毛2(2-C)の長さの通常0.7長以下（多くは0.6～0.7長）．櫛歯の形状は通常長歯と短歯差が明瞭.（九州以北に分布）

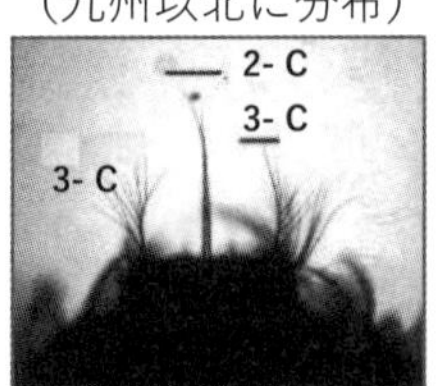

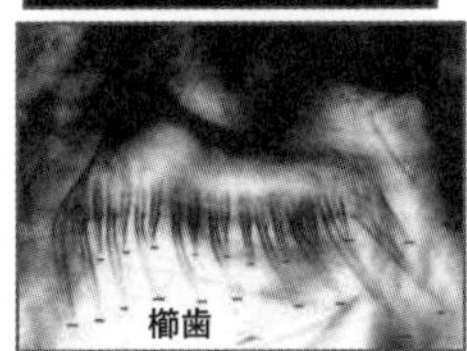

チョウセンハマダラカ
An. (Ano.) koreicus

頭毛3(3-C)は頭毛2(2-C)の長さの通常0.7長以上（多くは0.8～0.9長）．櫛歯の形状は通常長歯と短歯差が不明瞭.（沖縄本島以南に分布）

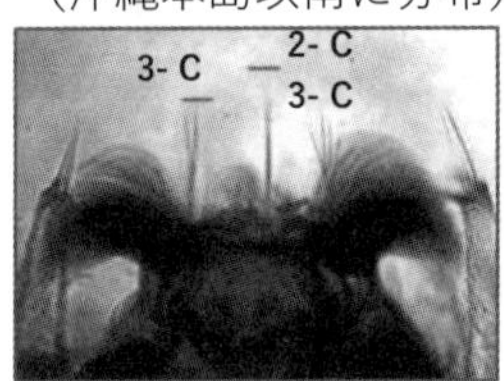

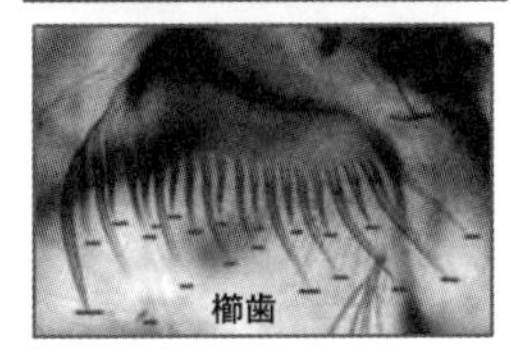

オオハマハマダラカ
An. (Ano.) saperoi

頭毛3(3-C)は10以上に分岐する。第Ⅱ～Ⅶ腹節毛０（0-Ⅱ～0-Ⅶ）は分岐し目立つ.

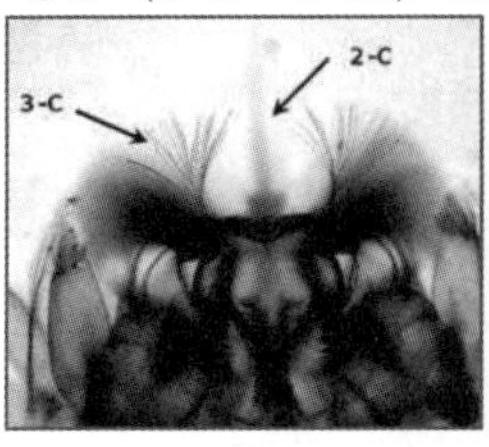

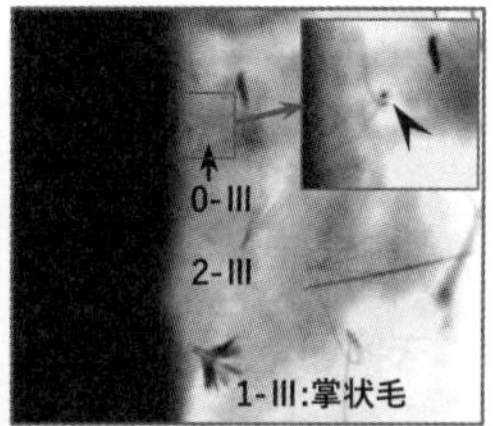

シナハマダラカ*An. (Ano.) sinensis*
オオツルハマダラカ*An. (Ano.) lesteri*
エンガルハマダラカ*An. (Ano.) engarensis*
エセシナハマダラカ*An. (Ano.) sineroides*
ヤツシロハマダラカ*An. (Ano.) yatsushiroensis*
（蛹での同定が正確ではあるが，下記により推定は可能）

通常触角毛1(1-A)は2～8本に分岐.

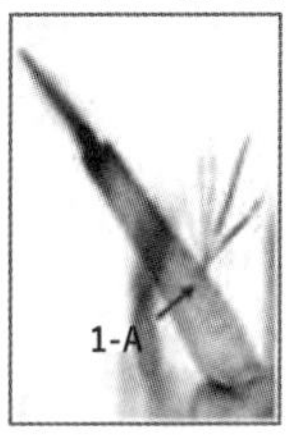

通常触角毛1(1-A)は8～11本に分岐.

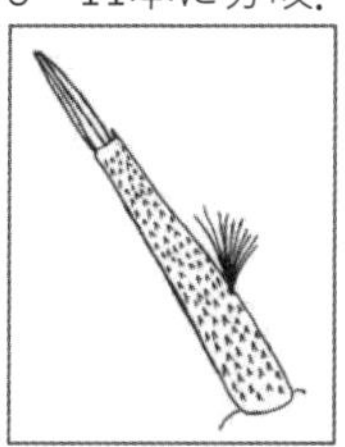

ヤツシロハマダラカ
An. (Ano.) yatsushroensis
（海岸近くに生息，長期間採集報告はない）

掌状毛は比較的大きく，各葉は細めで，先端はかなり細長く伸びる．各葉の色調はほぼ暗色で基部が淡色のものや先端側1/3が淡色のものあり.
(掌状毛が淡色のものあり注意を要する)

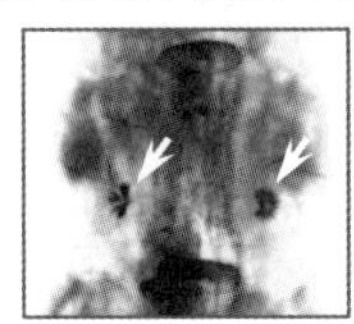

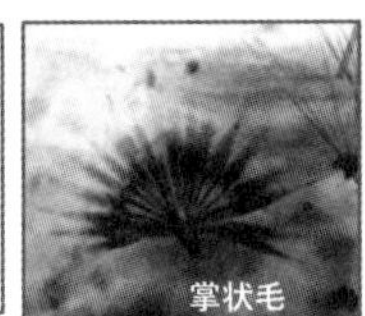

掌状毛は比較的小さく，各葉は太めで，先端はあまり細長く伸びない．各葉の色調にバラツキがあり，一様ではない.
(掌状毛が淡色のものあり注意を要する)

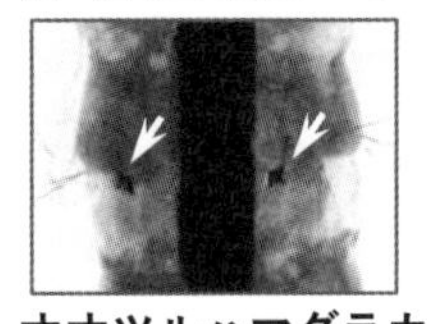

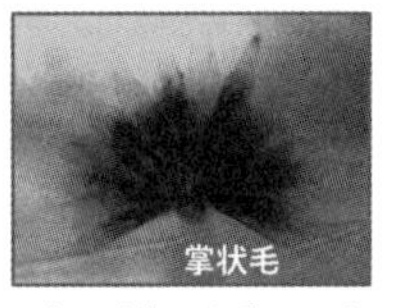

オオツルハマダラカ ***An. (Ano.) lesteri***
（沖縄県と北海道を除く地域ではまれな種となる）

通常頭毛3(3-C)は18～34本に分岐し，頭毛9(9-C)は7～12本に分岐する.

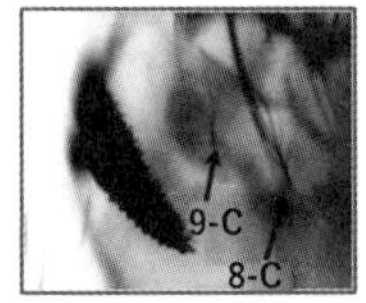

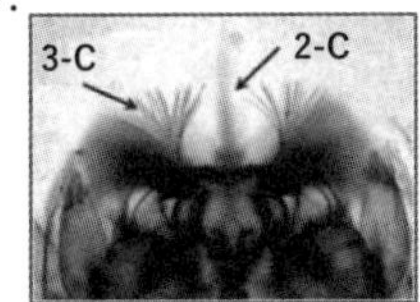

エセシナハマダラカ ***An. (Ano.) sineroides***

通常頭毛3(3-C)は26～89本に分岐し頭毛9(9-C)は.は5～9本に分岐する.

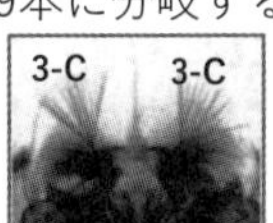

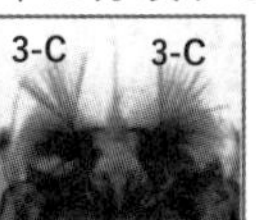

＊日本全土に生息.
＊＊北海道に生息，道南を除きハマダラカ属の中で最多.

シナハマダラカ ***An. (Ano.) sinensis ****
エンガルハマダラカ ***An. (Ano.) engarensis *****

A（3ページより）
イエカ属 *Culex*，カクイカ属 *Lutzia*

口刷毛は太い刷毛状．呼吸管長は鞍板長と同じか短い．第Ⅱ腹節毛7(7-Ⅱ)は第Ⅰ腹節毛7(7-Ⅰ)と同様顕著．

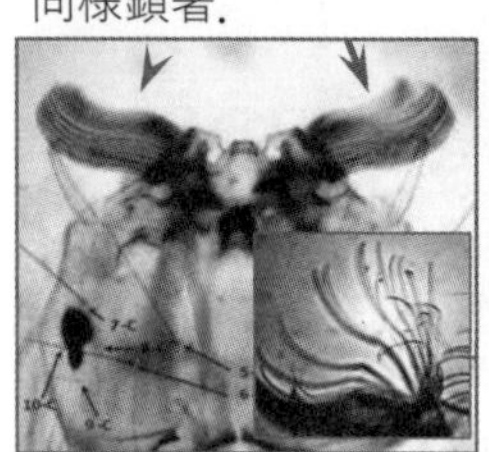

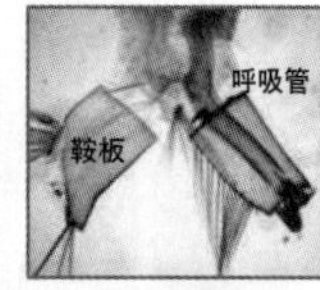

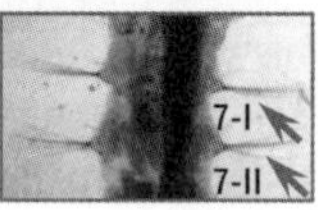

カクイカ属 *Lutzia*（旧カクイカ亜属 *Lutzia*）

呼吸管棘(PT)は呼吸管の先端近くまである．

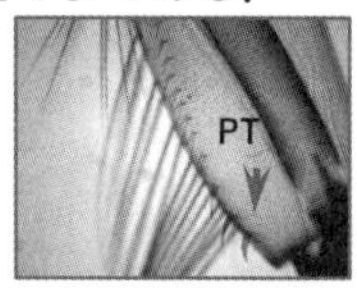

カクイカ亜属 *Metaltzia*
トラフカクイカ
Lu. (Mlt.) vorax
サキジロカクイカ
Lu. (Mlt.) fuscana

呼吸管棘(PT)は呼吸管の中ほどまで存在．

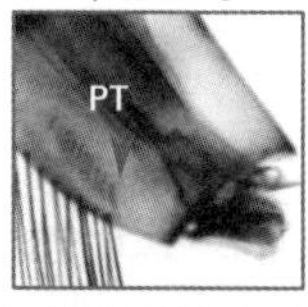

オガサワラカクイカ亜属
Insulalutzia
シノナガカクイカ
Lu. (Ilt.) shinonagai

高齢幼虫の両者の鑑別は困難である．若齢は生体においては容易であるが液浸保存で脱色したものは困難．

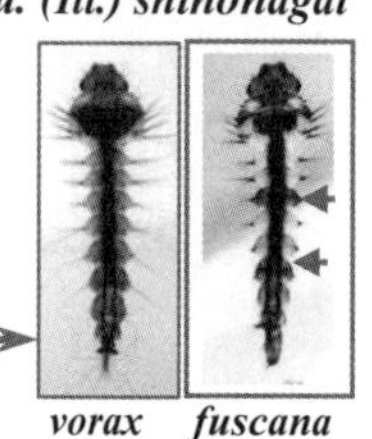

口刷毛は多数の微細毛．呼吸管長は鞍板長より長く，呼吸管末端部に呼吸管棘(PT)を欠く．第Ⅱ腹節毛7(7-Ⅱ)は第Ⅰ腹節毛7(7-Ⅰ)より短くて細い．

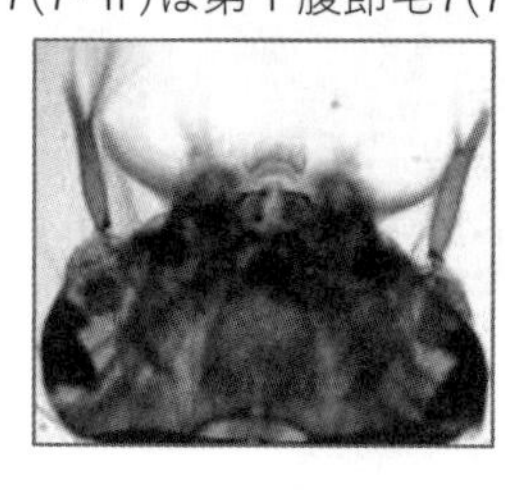

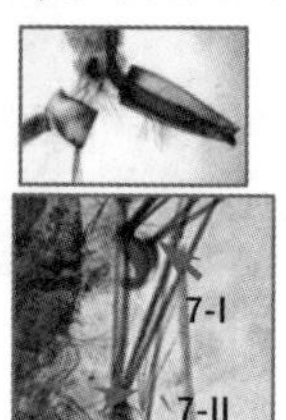

前胸毛3(3-P)は単条で，前胸毛1(1-P)とほぼ同太，同長．

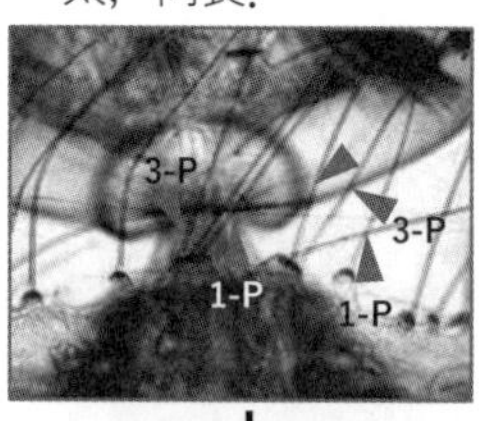

前胸毛3(3-3-P)は単条又は分岐し，前胸毛1(1-P)より細くてはっきりと短い．

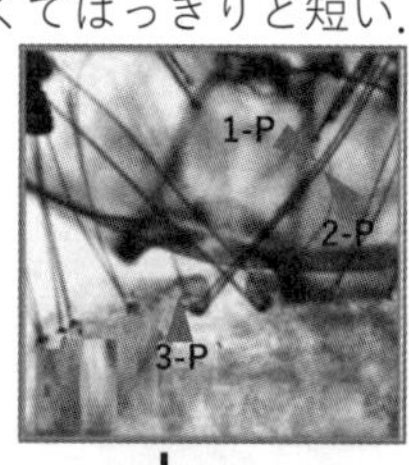

鞍板に明瞭な棘なし．

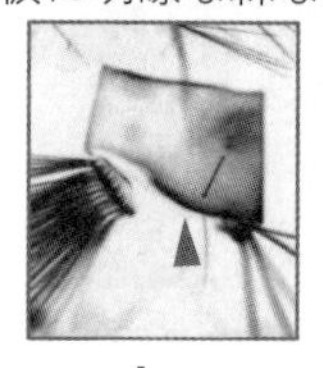

鞍板に明瞭な棘あり．

呼吸管毛1(1-S)は対をなさず．ジグザグにほぼ1列に生じる．

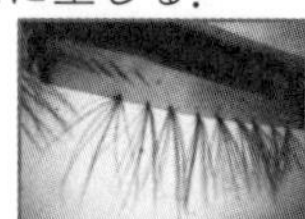

シオカ亜属 *Barraudius*

頭毛5(5-C)は頭毛6(6-C)の後方やや中央よりに生じ，通常2～3分岐．

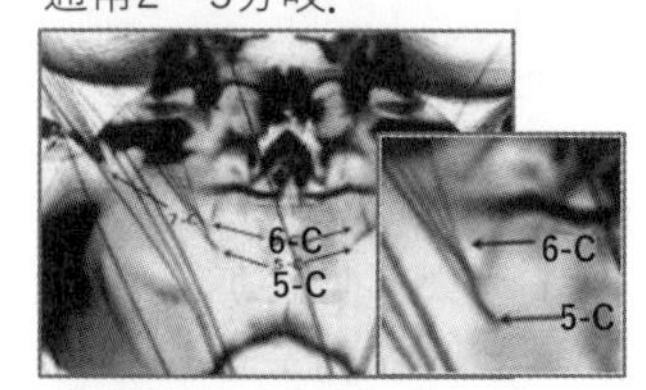

イナトミシオカ
Cx. (Bar.) inatomii

呼吸管毛1(1-S)は対をなし，側面にも呼吸管毛1(1-S)を生じる．

（10ページAへ）

腹面刷毛状毛(格子上毛)4(4-X)は5～7対(10束以上)時々対をなさない剛毛を格子の前方に生じる．鞍板に明瞭な棘なし．

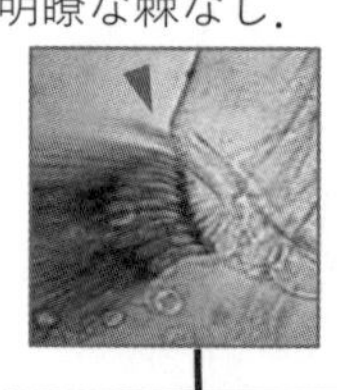

前胸毛4(4-P)は前胸毛3(3-P)より太くて長い．頭毛5(5-C)は頭毛7(7-C)より長い．

（8ページAへ）

前胸毛4(4-P)は前胸毛3(3-P)より細くて短い．頭毛5(5-C)は頭毛7(7-C)より短い．

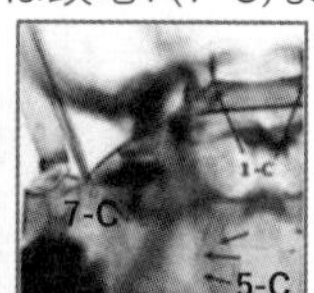

クロウスカ亜属 *Eumelanomyia* の一部

（9ページBへ）

腹面刷毛状毛(格子上毛)4(4-X)は4対(8束)のみ．鞍板に明瞭な棘あり．

クシヒゲカ亜属
Culiciomyia

（13ページAへ）

A（7ページより）

呼吸管棘(PT)は個々に少なくとも4側歯を有し，しばしば約10側歯を有する．

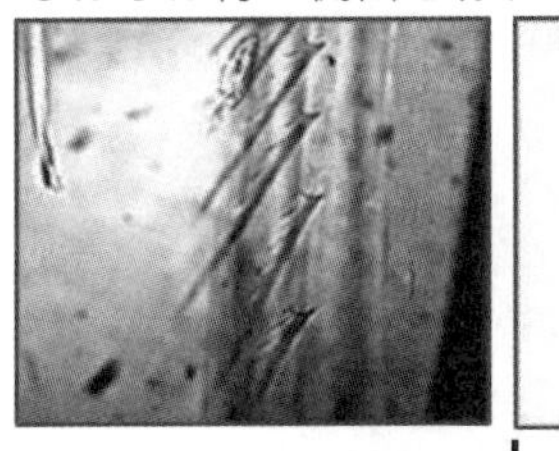
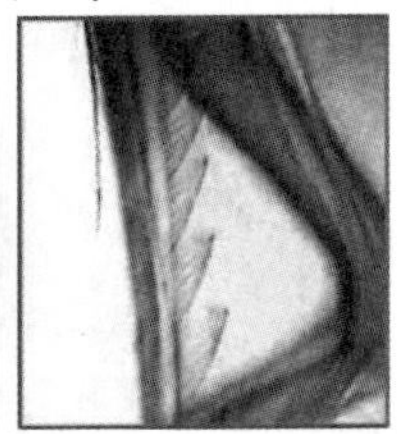

触角毛2,3(2-,3-A)は触角の先端から生じる．

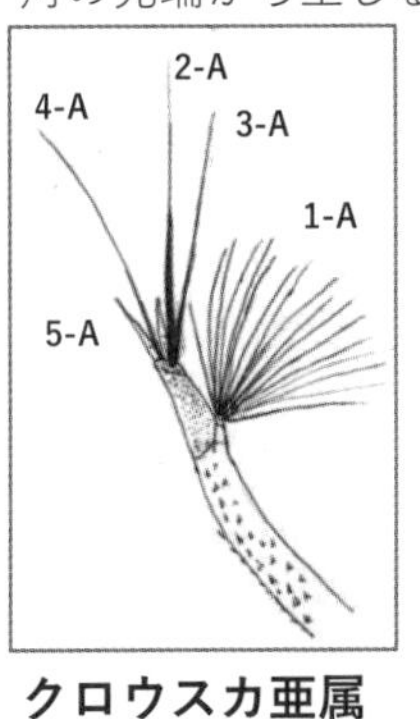

クロウスカ亜属
Eumelanomyia
の一部
（9ページBへ）

触角毛2,3(2-,3-A)は触角の亜先端から生じる．

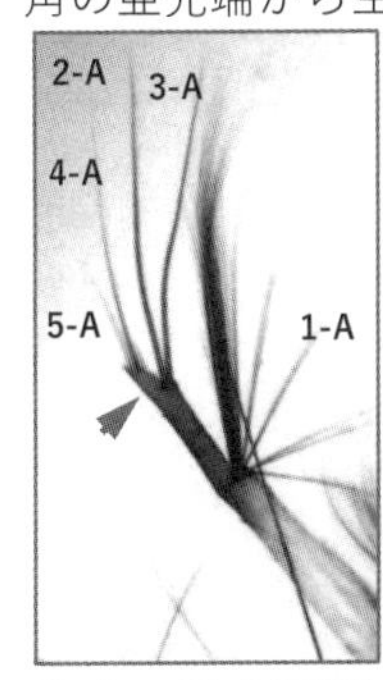

ツノフサカ亜属
Lophoceraomyia

側鱗(CS)はしゃもじ型であるが後方列のものは通常先端棘が他の側棘より太くて尖る．

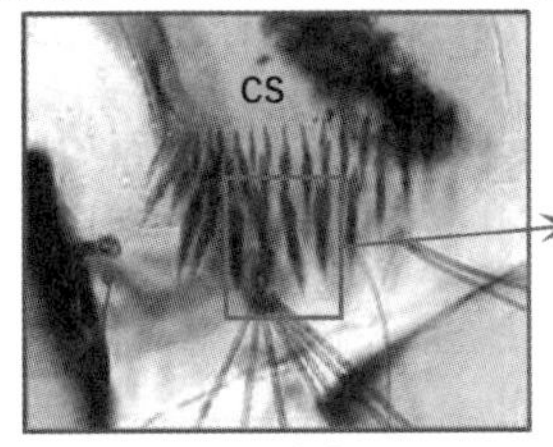

クロツノフサカ *Cx. (Lop.) bicornutus*

胸部外皮は全体に微棘状を呈し，前胸毛3(3-P)は前胸毛2(2-P)に比べ非常に短く3〜7分岐する．

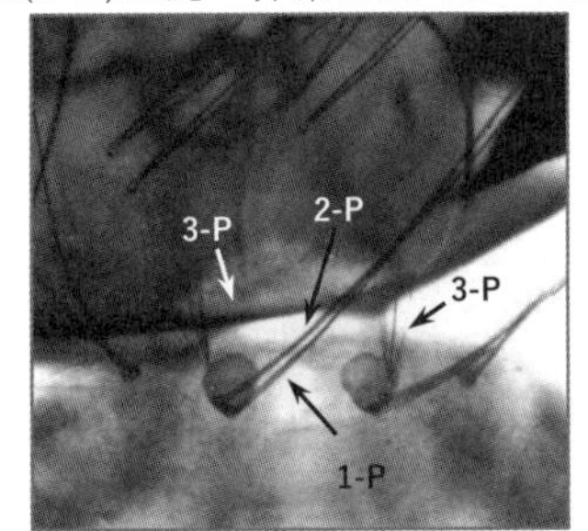

アカツノフサカ *Cx. (Lop.) rubithoracis*

呼吸管棘(PT)は個々に多くとも3側歯（外国産種では超えるものあり）で，通常は1〜2側歯．

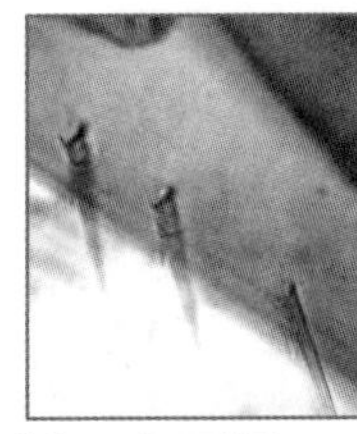

エゾウスカ亜属 *Neoculex*

頭毛4(4-C)は細く2分岐（稀に単条）で頭毛6(6-C)のやや上方，かなり正中寄りに生じる．頭毛5(5-C)は通常2分岐（まれに3分岐）で頭毛6（6-C)後方やや正中寄りに生じる．6-Cは常に2分岐で5-Cとほぼ同長．

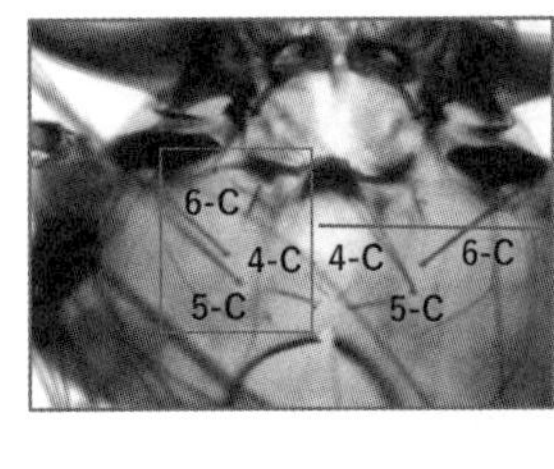

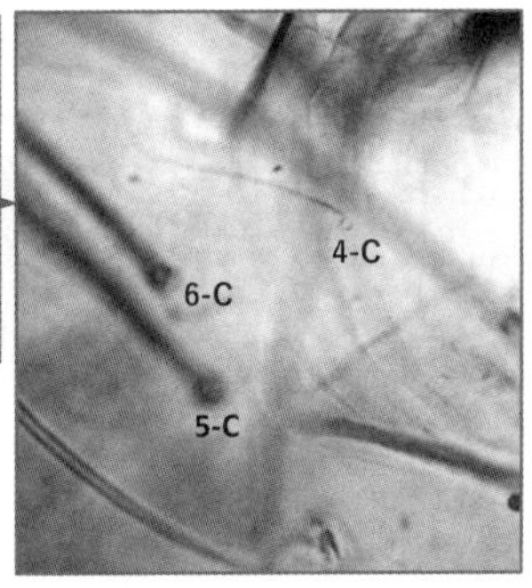

エゾウスカ *Cx. (Ncx.) rubensis*

側鱗(CS)は前後列共しゃもじ型で先端棘は他の側棘と同じで太くならない．

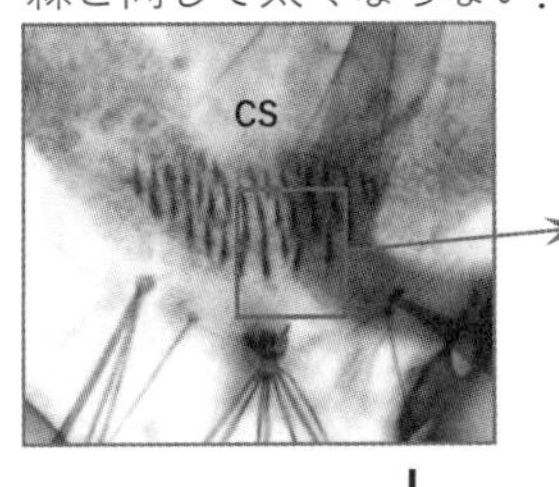

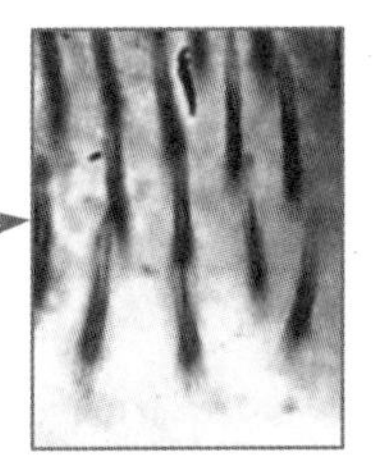

胸部外皮は滑らかで，前胸側部のみ微棘状を呈す．前胸毛3(3-P)は前胸毛2(2-P)に比べ非常に短くはなく1〜2分岐．

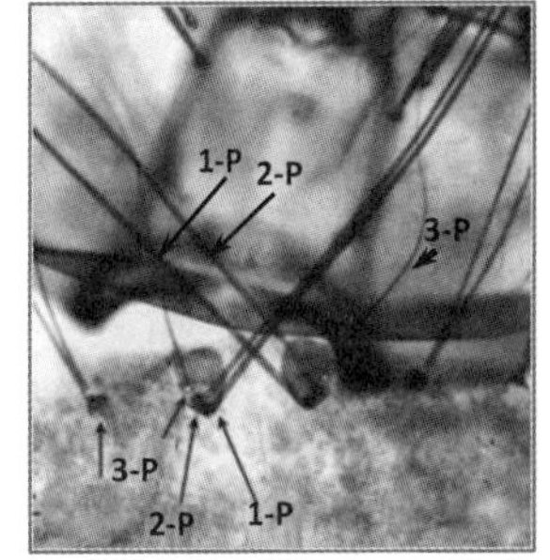

（9ページAへ）

A（8ページより）

B（7,8ページより）
クロウスカ亜属 *Eumelanomyia*

頭毛5(5-C)は通常3分岐（時折片側だけ2分岐又は4分岐）．前胸毛14(14-P)は単条．第VIII腹節毛2(2-VIII)は通常2分岐．

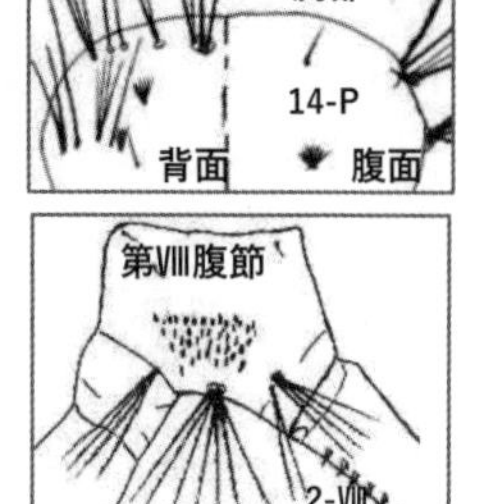

ハラオビツノフサカ *Cx. (Lop.) cinctellus*

頭毛5(5-C)は単条か2分岐（時折片側だけ2分岐）．前胸毛14(14-P)は2〜4分岐．第VIII腹節毛2(2-VIII)は通常単条（ごく稀に2分岐）．

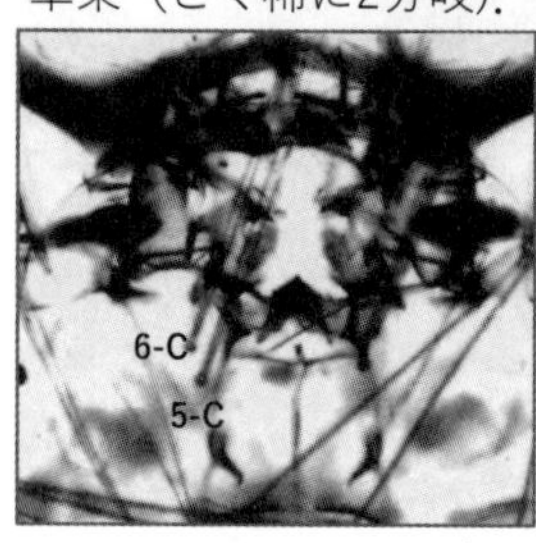

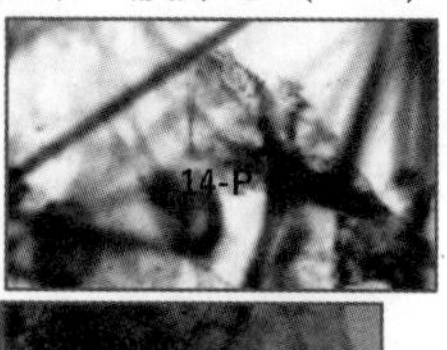

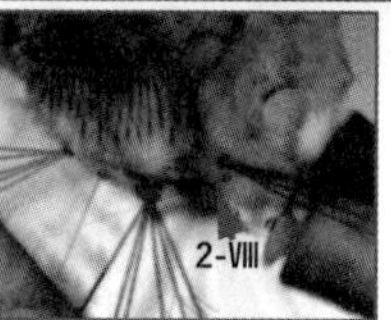

頭毛6(6-C)はほとんど常に2分岐．頭毛5(5-C)は常に2分岐．第VII腹節毛1(1-VII)は6〜7分岐．呼吸管比は8〜11．

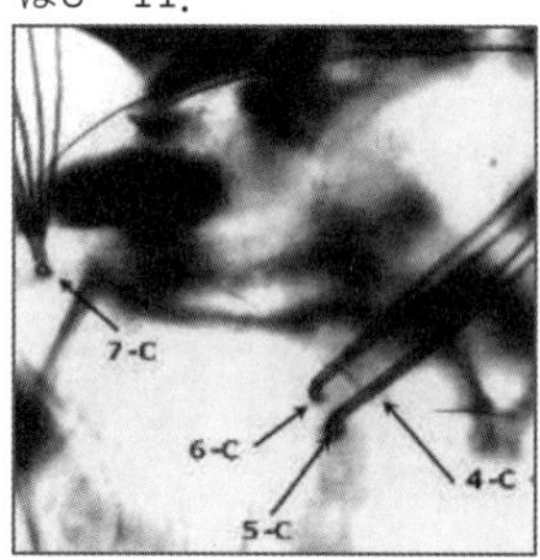

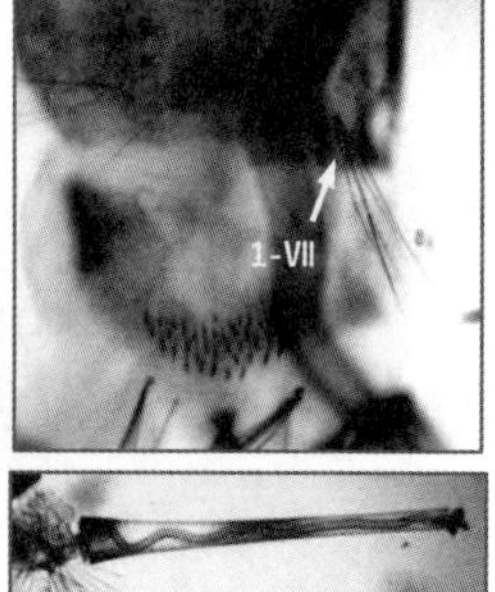

フトシマツノフサカ *Cx. (Lop.) infantulus*

頭毛6(6-C)は単条．頭毛5(5-C)は通常単条．第VII腹節毛1(1-VII)は2分岐．呼吸管比は6〜7.5．

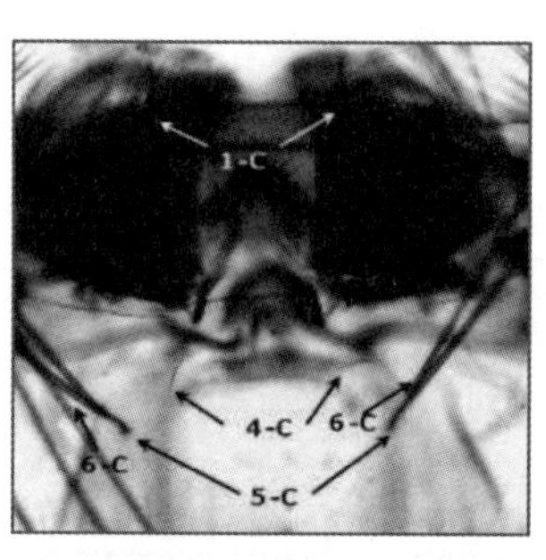

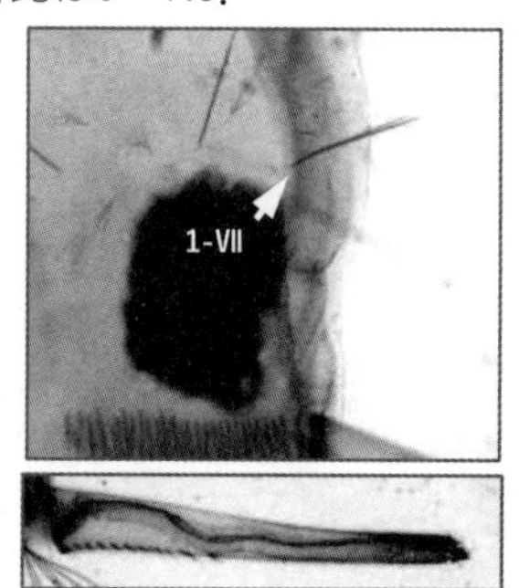

カニアナツノフサカ *Cx. (Lop.) tuberis*

呼吸管毛1(1-S)はそれぞれの位置の呼吸管幅より短い．呼吸管は褐色で呼吸管比は10以上と長い．呼吸管棘(PT)は基部側1/5以内に生じる．

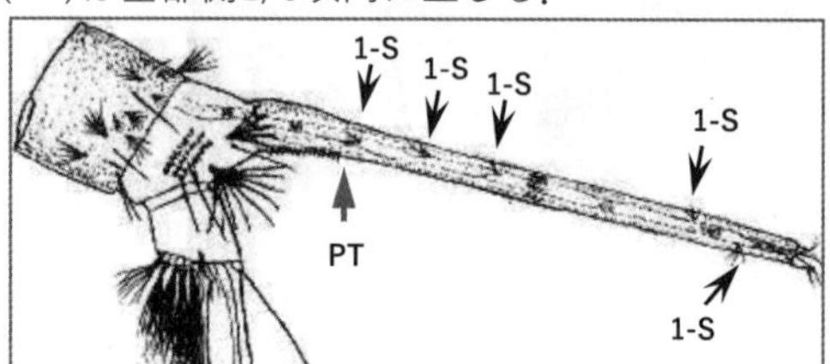

カギヒゲクロウスカ *Cx. (Eum.) brevipalpis*

呼吸管毛1(1-S)はそれぞれの位置の呼吸管幅より長い．呼吸管は褐色で呼吸管比は8以下と短い．呼吸管棘(PT)は基部側1/5を越えて生じる．

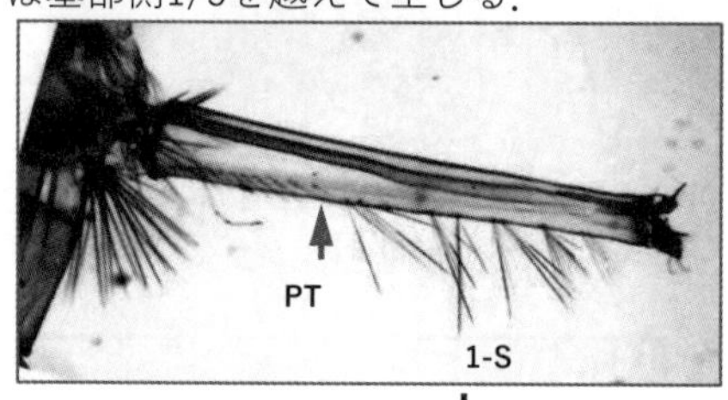

触角は淡色．背面刷毛状毛2(2-X)は2〜5分岐．第III〜VI腹節毛6(6-III〜VI)は3分岐．

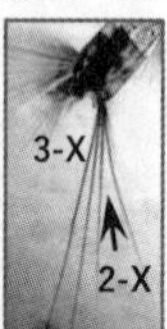

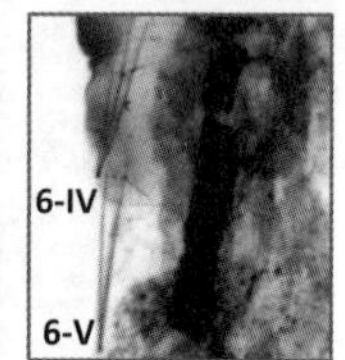

コガタクロウスカ

Cx. (Eum.) hayashii hayashii コガタクロウスカ*

Cx. (Eum.) hayashii ryukyuanus リュウキュウクロウスカ**

*屋久島以北に生息．　**奄美大島以南に生息．

触角は暗色．背面刷毛状毛2(2-X)は通常単条．第III，IV，V腹節毛6(6-III，IV，V)は2分岐．

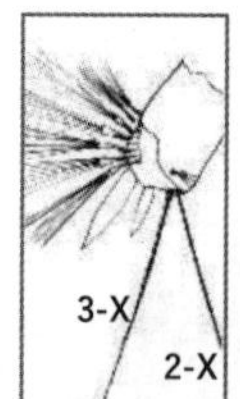

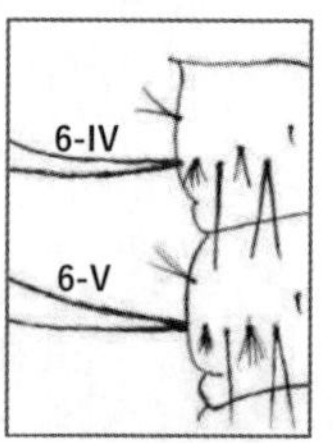

オキナワクロウスカ *Cx. (Eum.) okinawae*

A（7ページより）

イエカ亜属 *Culex*, カラツイエカ亜属 *Oculeomyia* 及びオガサワライエカ亜属 *Sirivanakarnius*

側鱗(CS)はしゃもじ状で側面から先端まで見事な側棘で縁取られる。

CS

イエカ亜属 *Culex*（一部）

側鱗(CS)は棘状で先端の棘は側面の棘より太くて大きい.

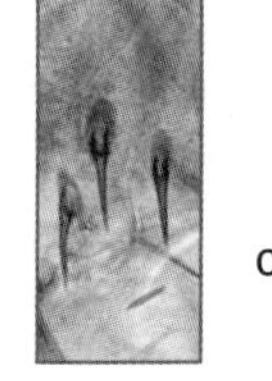

CS

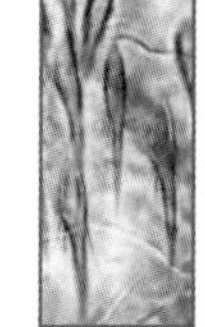

（11ページBへ）

頭毛1(1-C)は細くて先端は糸状.

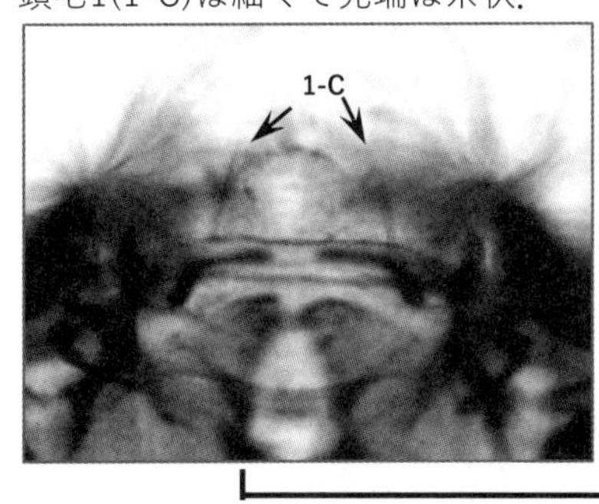

頭毛1(1-C)は暗色で短い.

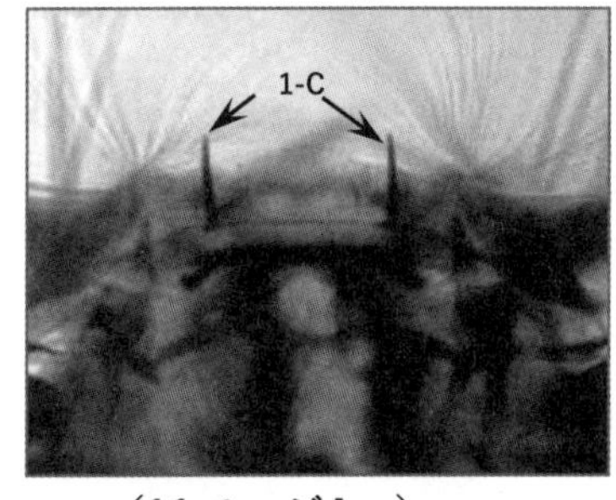

（11ページAへ）

頭毛5,6(5-,6-C)は2〜3分岐．呼吸管毛1(1-S)は非常に弱々しく，それぞれの位置の呼吸管幅より短い．呼吸管比は5を越える．

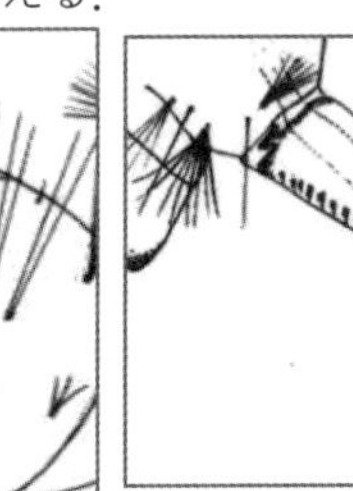

1-S
1-S
1-S

オビナシイエカ *Cx. (Cux.) fuscocephala*

頭毛5,6(5-,6-C)は4〜7分岐．呼吸管毛1(1-S)はよく発達し，それぞれの位置の呼吸管幅より長いか又は呼吸管比が5未満.

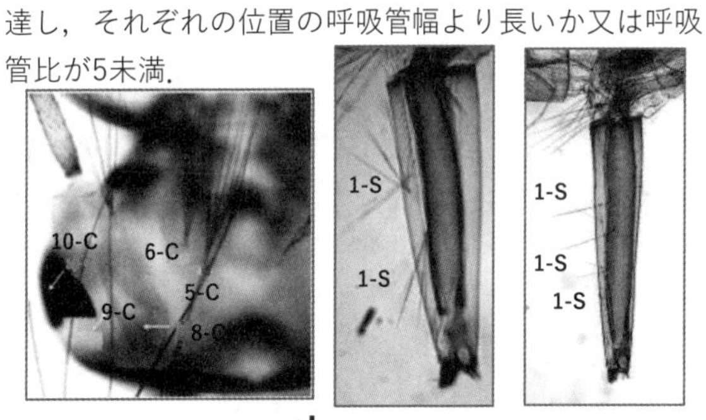
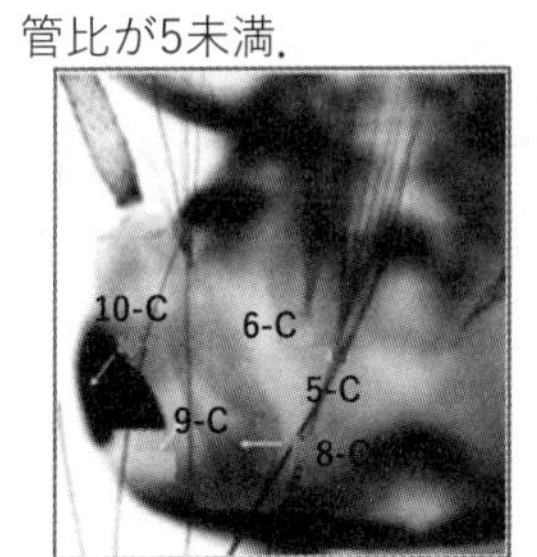

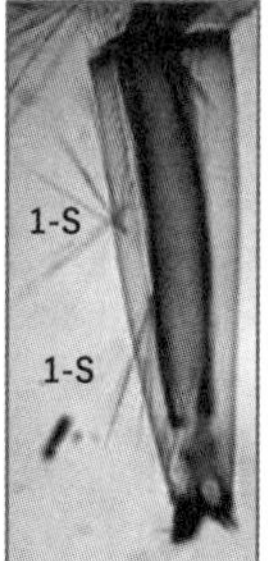

1-S
1-S
1-S

呼吸管比は5を越える．呼吸管毛1(1-S)は通常5対.
下唇板の歯は16〜20歯.

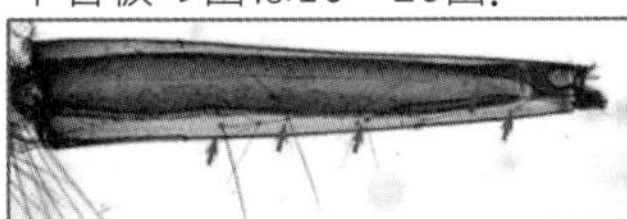

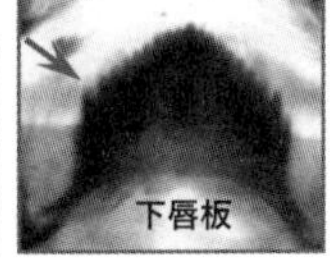

スジアシイエカ *Cx. (Cux.) vagans*

呼吸管比は5以下．呼吸管毛1(1-S)は通常4対.
下唇板の歯は20〜25歯.

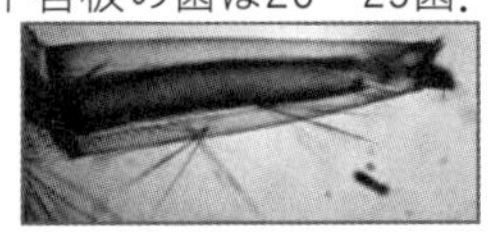

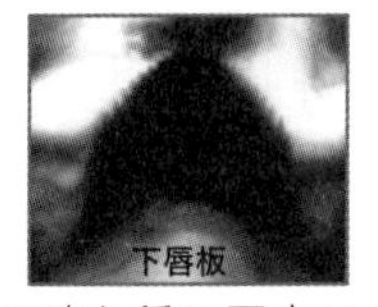

（これ以上は推定に止める．正確な種の同定は♂の外部生殖器による）

呼吸管は通常中央付近が幾分膨れる．呼吸管比は通常3〜4.2．呼吸管毛(1-S)は4対で5〜8分岐.（通常6分岐分岐以上）.

ネッタイイエカ *Cx. (Cux.) quinquefasciatus*

呼吸管は通常中央付近は膨れない．呼吸管比は通常3.5〜5.6．呼吸管毛(1-S)は4〜5対(通常4対，稀に5対)で2〜6分岐．（通常5分岐以下）.

アカイエカ *Cx. (Cux.) pipiens pallens*
チカイエカ *Cx. (Cux.) pipiens* form *molestus*

A（10ページより）

B（10ページより）

前胸毛4(4-P)は単条．呼吸管毛(1-S)はほとんど腹面上に生え，基部側のほとんどは6～13分岐．

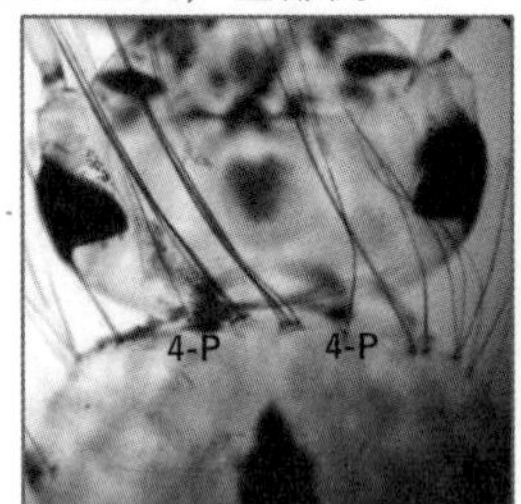

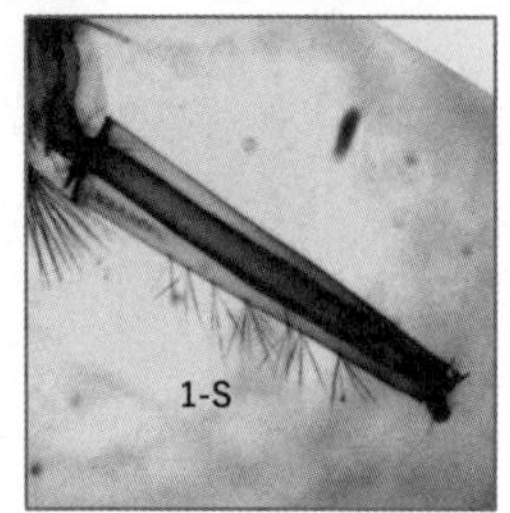

オガサワライエカ亜属 ***Sirivanakarnius***
オガサワライエカ ***Cx. (Sir.) boninensis***

前胸毛4(4-P)は2分岐．呼吸管毛(1-S)は腹面近くに生え，通常2～5分岐．

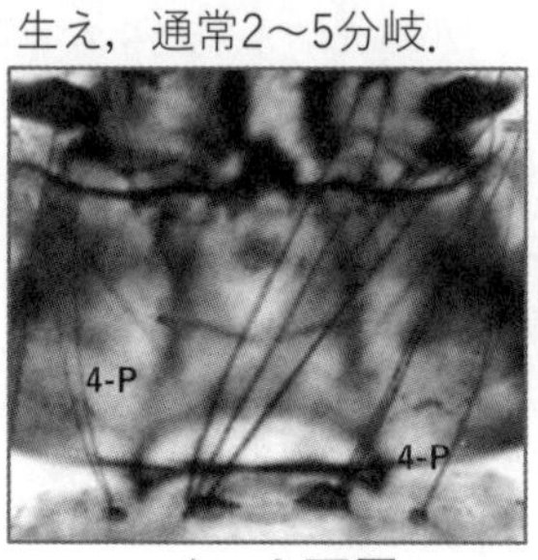

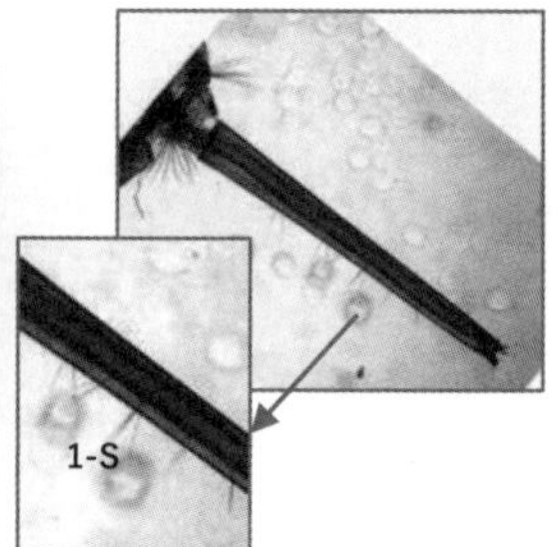

イエカ亜属 ***Culex*** (一部)

頭毛5(5-C)は2～4分岐．頭毛6(6-C)と第I腹節毛7(7-I)は2分岐．呼吸管比は7.0～9.0.

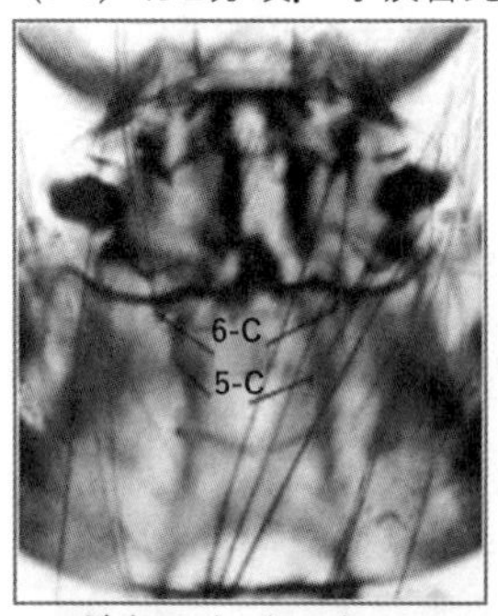

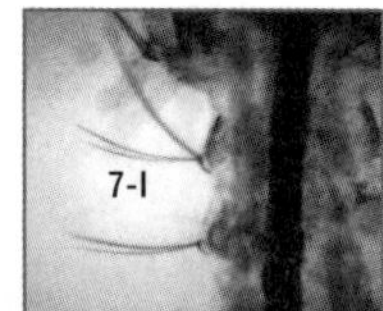

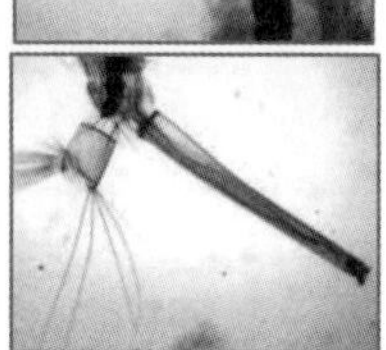

コガタアカイエカ ***Cx. (Cux.) tritaeniorhynchus***

頭毛5(5-C)は5～7分岐．頭毛6(6-C)は4～7分岐．第I腹節毛7(7-I)は単条．呼吸管比は5.2～6.2.

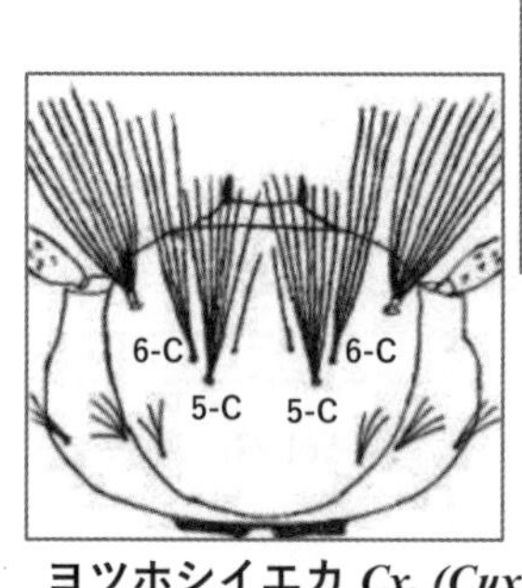

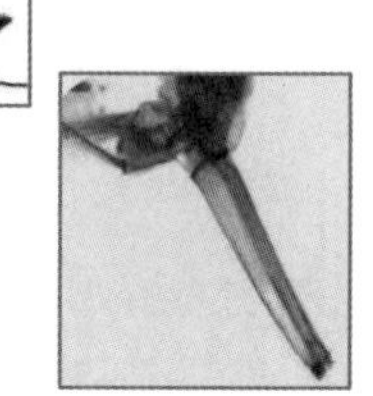

ヨツホシイエカ ***Cx. (Cux.) sitiens***

側鱗(CS)は通常15個未満でほぼ1～2列に並ぶ．呼吸管毛1(1-S)は腹面近くにある．

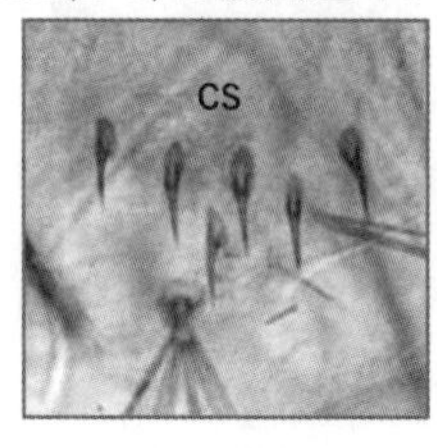

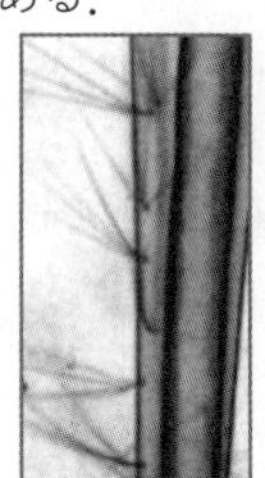

側鱗(CS)は通常16個以上で3～4列に並ぶ．呼吸管毛1(1-S)はほとんど腹面上にある．

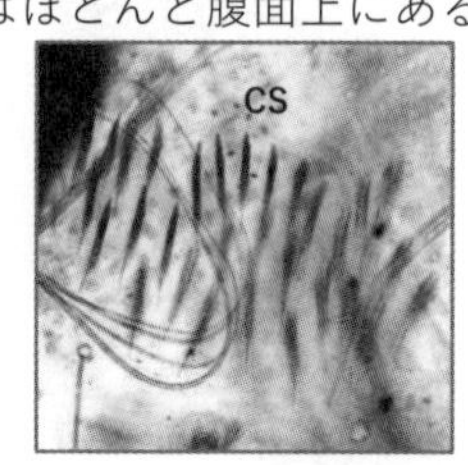

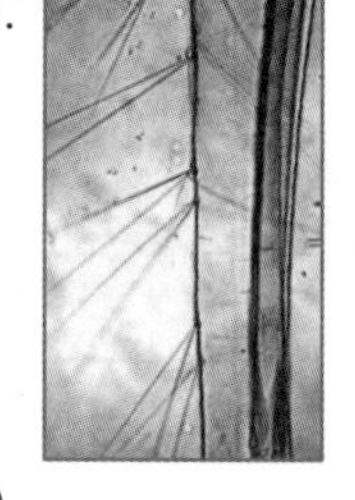

(12ページCへ)

呼吸管棘(PT)は少なくとも呼吸管の基部側2/10位置まで生える．

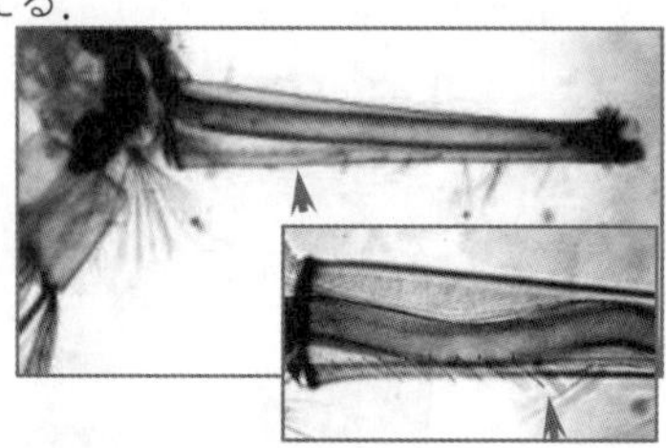

(12ページAへ)

呼吸管棘(PT)はないか痕跡的で多くとも呼吸管の基部側2/10位置までしか生えない．

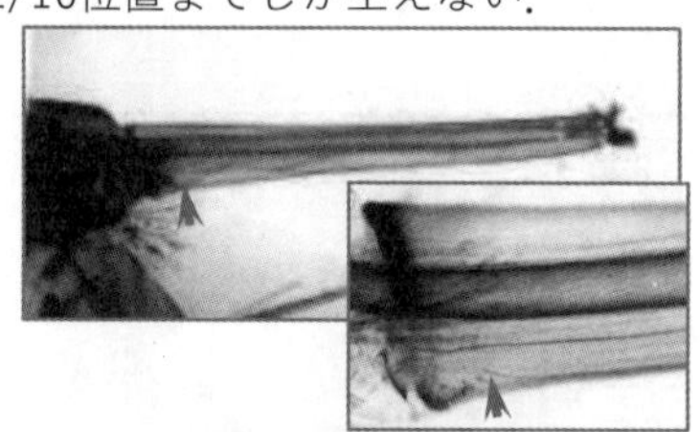

カラツイエカ亜属 ***Oculeomyia***

(12ページBへ)

A（11ページより）

呼吸管毛1(1-S)の長さは中程度で呼吸管幅の2倍以上はない．頭毛5(5-C)は3～4分岐(片側2分岐のものあり)．背面刷毛状毛2(2-X)は3～4分岐．

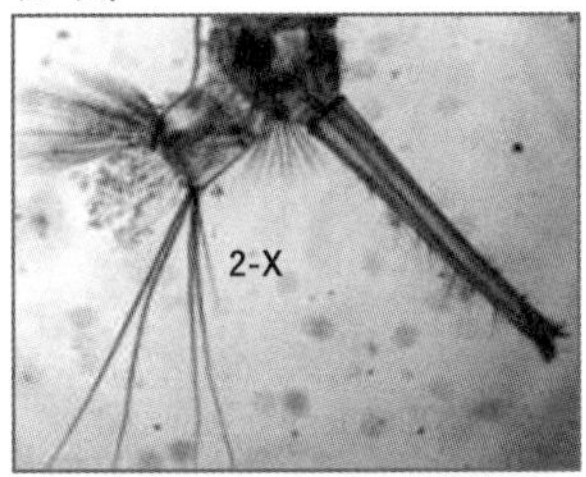

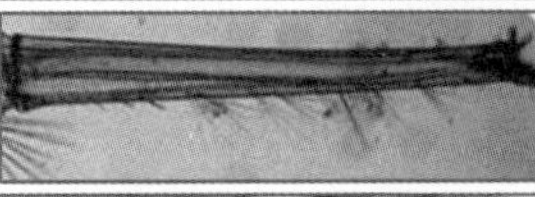

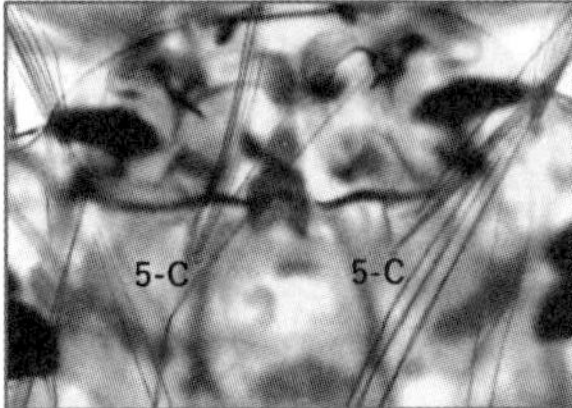

シロハシイエカ
Cx. (Cux.) pseudovishnui

呼吸管毛1(1-S)は長く呼吸管幅の2倍長を越える．頭毛5(5-C)は2分岐．背面刷毛状毛2(2-X)は単条．

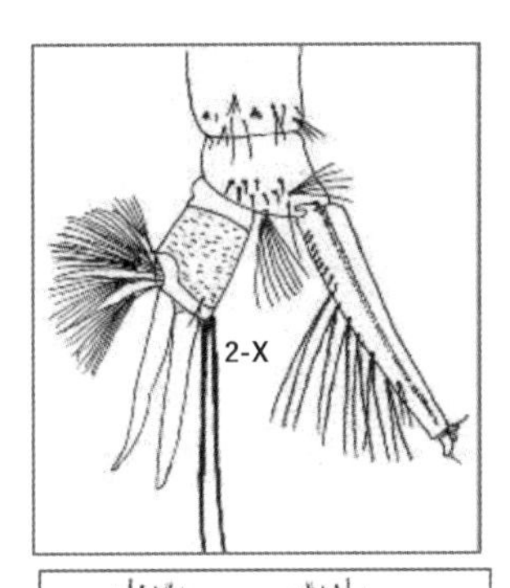

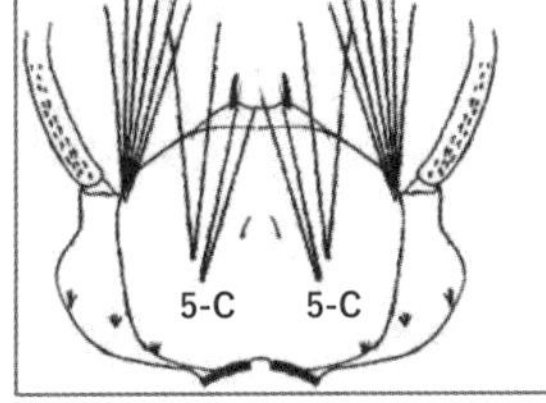

セシロイエカ
Cx. (Cux.) whitmorei

B（11ページより）

下唇板の歯ははっきりし19～27歯．頭毛4(4-C)は頭部前縁を越えて伸びる．前胸毛4(4-P)は短い．

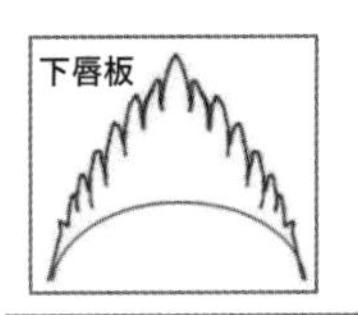

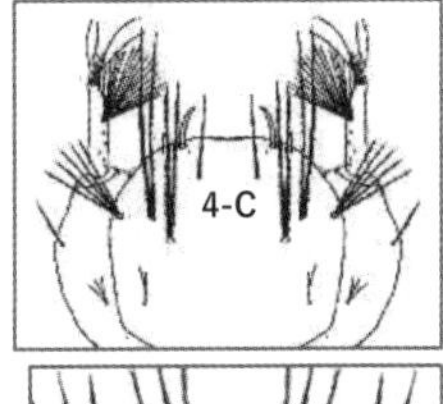

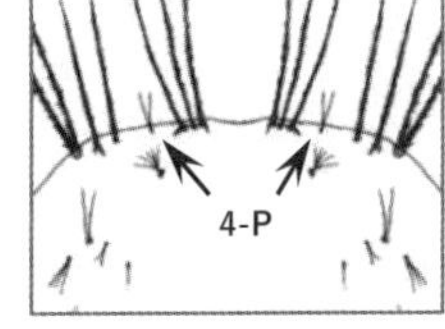

ミツホシイエカ
Cx. (Ocu.) sinensis

C（11ページより）

下唇板の歯は極めて細かく約100歯で目立たない．頭毛4(4-C)は頭部前縁を越えない．前胸毛4(4-P)は長い．

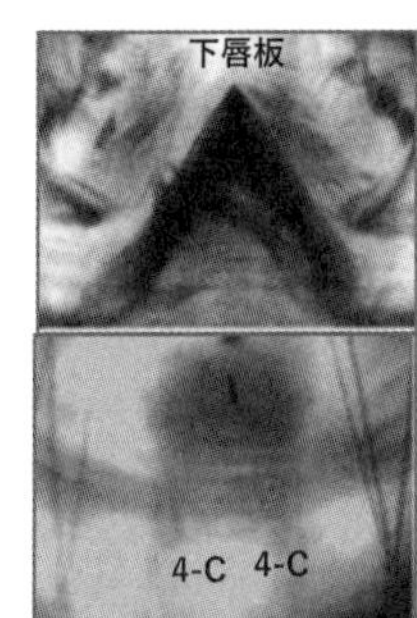

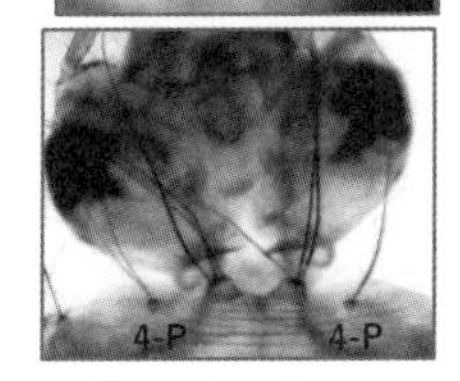

カラツイエカ
Cx. (Ocu.) bitaeniorhynchus

呼吸管先端側1/2に側歯を伴わない太い棘を1～4本有す．腹面の呼吸管毛1(1-S)は両側で6～8束．頭毛4(4-C)は3～6分岐．

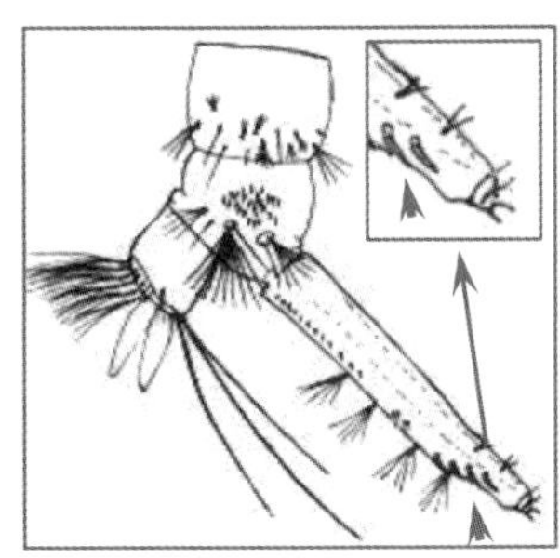

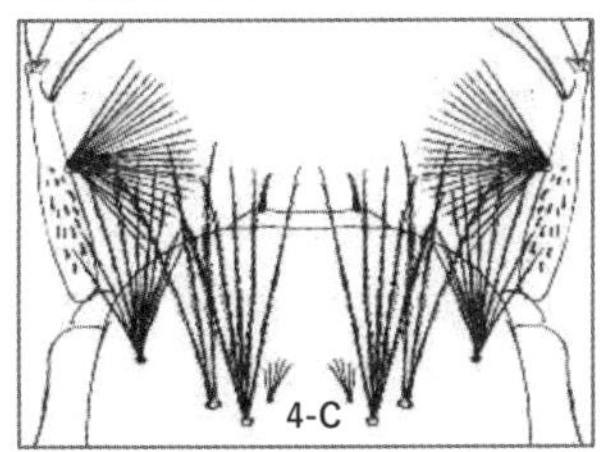

ジャクソンイエカ
Cx. (Cux.) jacksoni

呼吸管棘(PT)はすべて側歯を伴い，腹面の呼吸管毛1(1-S)は両側で9～14束．

前胸毛4(4-P)は通常2分岐．

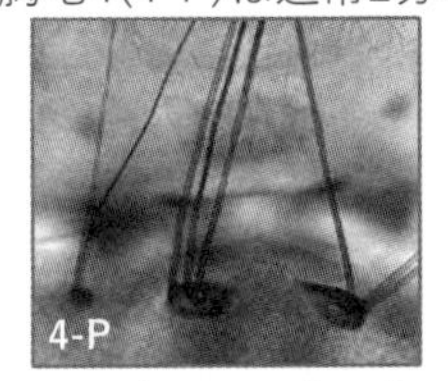

ニセシロハシイエカ
Cx. (Cux.) vishnui

前胸毛4(4-P)は通常単条．

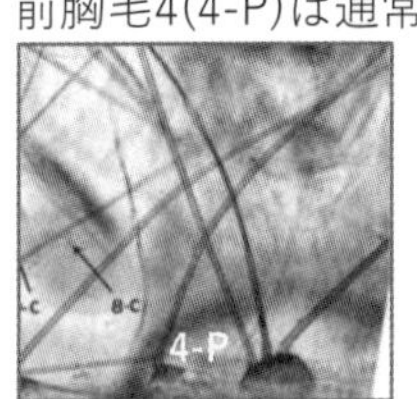

頭毛4(4-C)は3～5分岐．下唇板は三角形で前面に歯状突起が等しく並ぶ．先端の呼吸管毛2(2-S)は湾曲し先端側の呼吸管棘(PT)より長い．

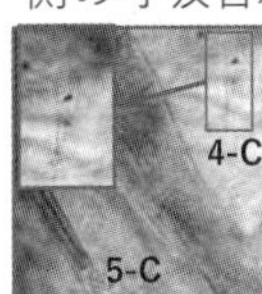

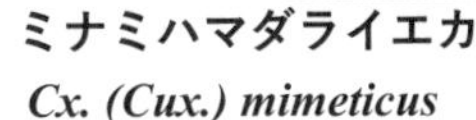

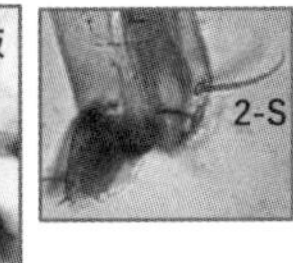

ミナミハマダライエカ
Cx. (Cux.) mimeticus

頭毛4(4-C)は1～2分岐．下唇板は五角形で前面の歯状突起は不揃に並ぶ．先端の呼吸管毛2(2-S)は真っすぐで先端側の呼吸管棘(PT)より短い．

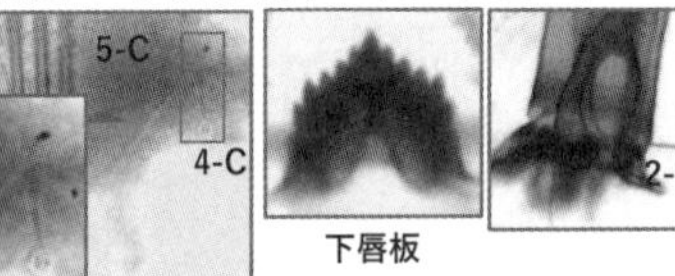

ハマダライエカ
Cx. (Cux.) orientalis

- 12 -

A（7ページより）
A（7ページより）

クシヒゲカ亜属 *Culiciomyia*

呼吸管に偽関節(硬化した細い帯)を伴い呼吸管比は9以上．頭毛5,6(5-,6-C)は通常3分岐．触角毛1(1-A)は触角の先端側約1/3に生える．

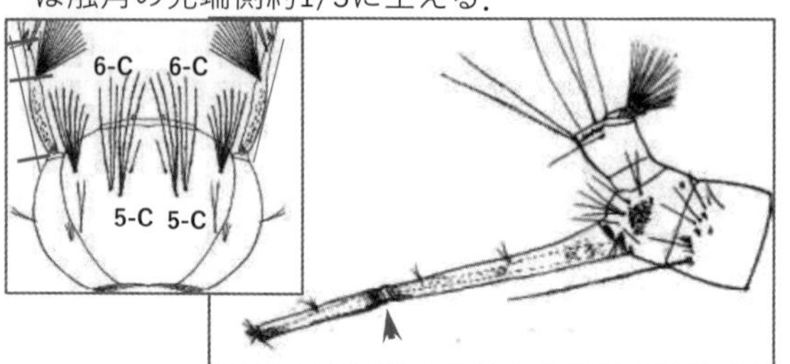

クロフクシヒゲカ *Cx. (Cui.) nigropunctatus*

呼吸管に偽関節は伴わず呼吸管比は9以下．

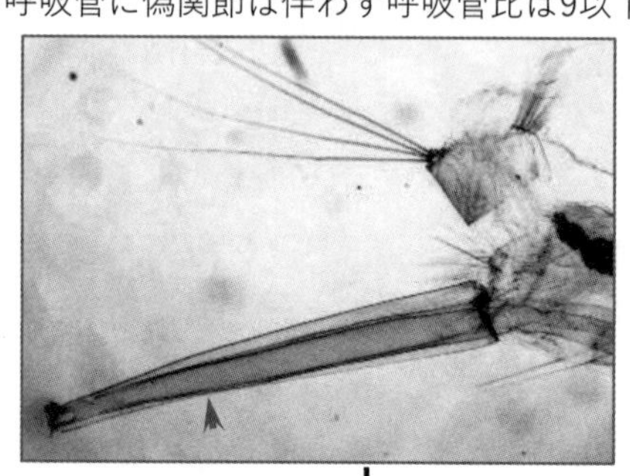

呼吸管は中央で膨らみ呼吸管棘(PT)は10以下．触角毛1(1-A)は通常触角の基部側約1/2に生える．

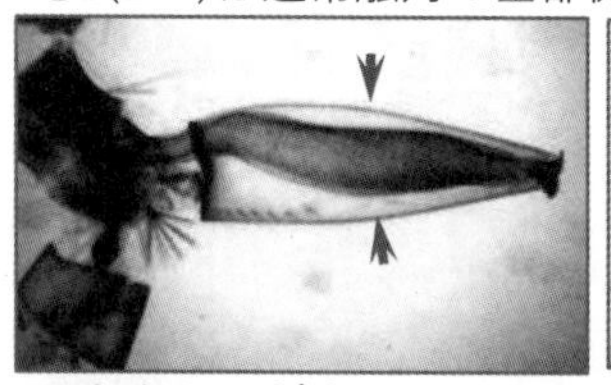

アカクシヒゲカ *Cx. (Cui.) pallidothorax*

呼吸管は中央で膨らまず，呼吸管棘(PT)は10以上．

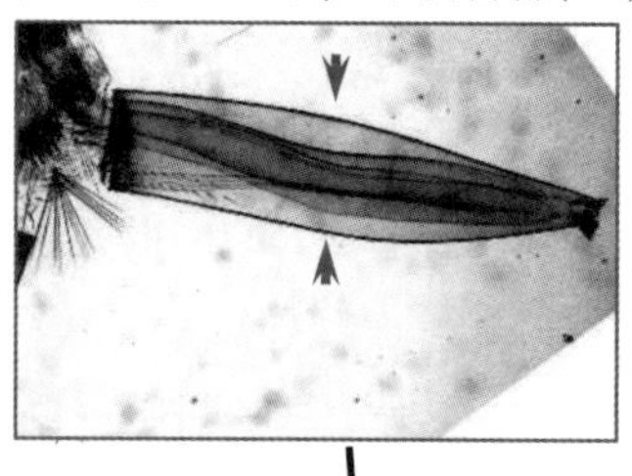

呼吸管毛1(1-S)は細く3対，触覚毛1(1-S)は通常触角の基部側約1/2に生える．

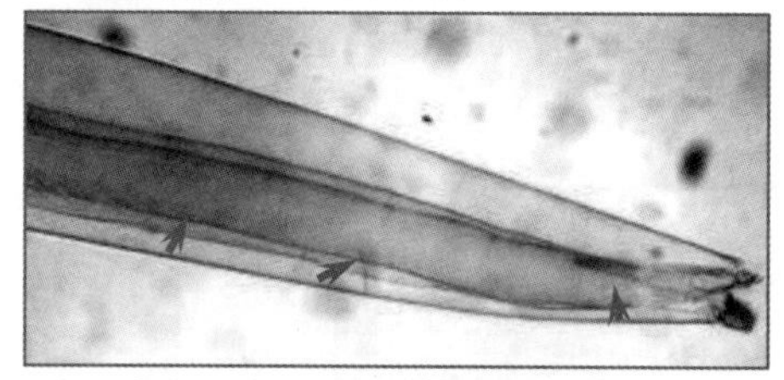

リュウキュウクシヒゲカ *Cx. (Cui.) ryukyensis*

呼吸管毛1(1-S)は4対以上．

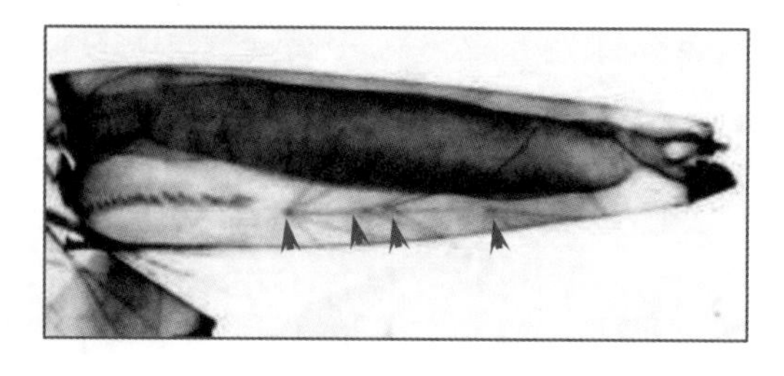

鞍板にくびれがない．呼吸管比は6.1以上．呼吸管毛1(1-S)は弱々しく，長さは呼吸管幅より短くて通常は単条．

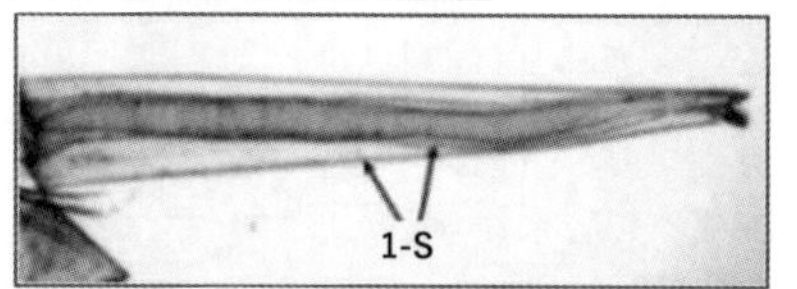

キョウトクシヒゲカ *Cx. (Cui.) kyotoensis*

鞍板は腹面刷毛状毛近くでくびれる．呼吸管比は5.8以下．呼吸管毛1(1-S)は普通で，長さは呼吸管幅とほぼ同じで通常3分岐以上に分岐．

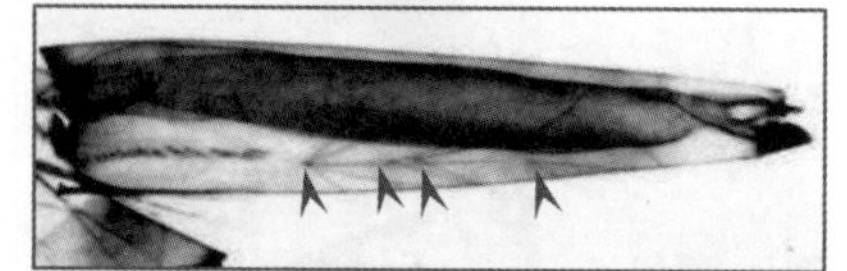

ヤマトクシヒゲカ *Cx. (Cui.) sasai*

A（4ページより）
ヤブカ属 *Aedes* 及びフトオヤブカ属 *Verrallina*

呼吸管基部の呼吸管棘列延長上位置に突起なし．
小腮軸節基部は切断される．

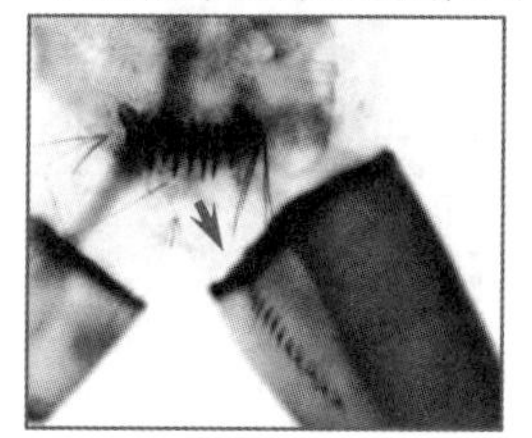
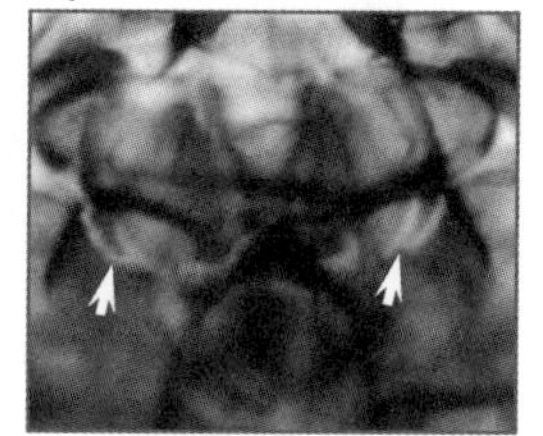

シマカ亜属 *Stegomyia* （16ページAへ）

呼吸管基部の呼吸管棘列延長上位置に突起あり．
小腮軸節基部は切断されずつながる．

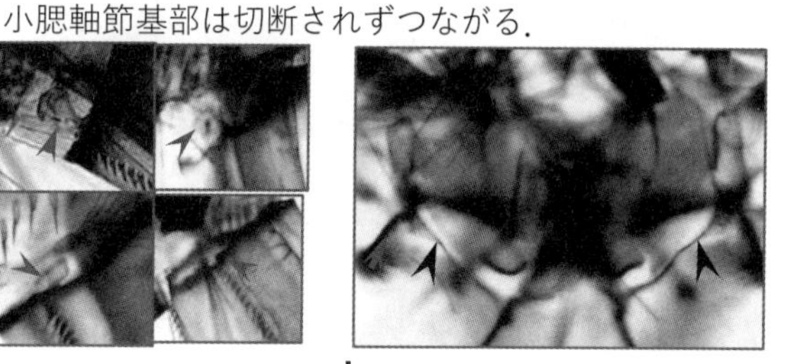

第Ⅰ腹節毛12(12-I)は存在する　　第Ⅰ腹節毛12(12-I)は存在しない．

(腹面の毛は構造上観察しづらく、特に12-Iの無い種は多くの場合10-I以外の毛は微細でかつ観察しづらい．12-Iは9-Iの比較的近い位置に生えるので9-Iから正中にかけて探すと良い)

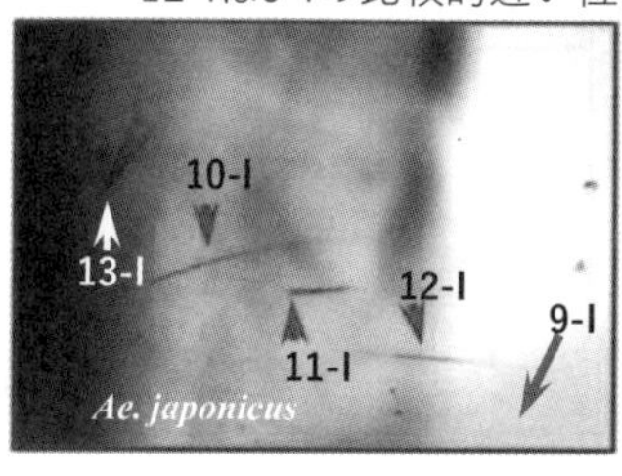

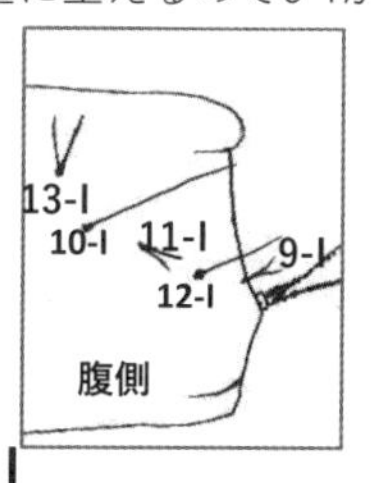

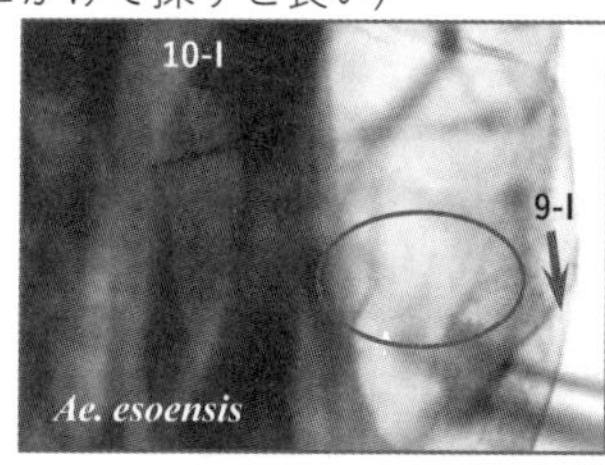

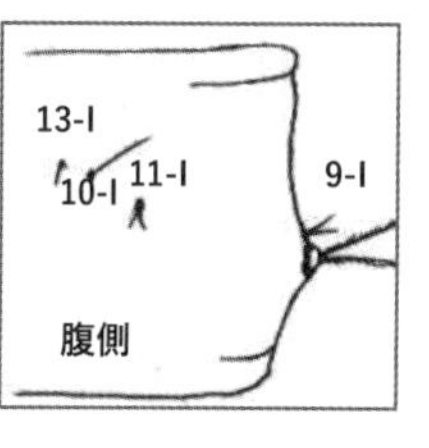

（15ページAへ）

頭毛8(8-C)は頭毛9,10(9-,10-C)の2倍長以上．尾葉は非常に短くて丸い．

カニアナヤブカ亜属 *Geosukusea*

頭毛5(5-C)は通常2分岐．頭毛6(6-C)は通常は単条．頭毛7(7-C)は通常10～11分岐．背面刷毛状毛2(2-X)は10～16分岐．

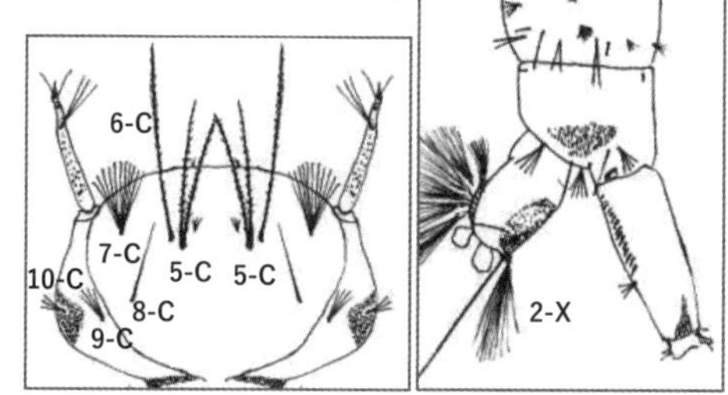

カニアナヤブカ *Ae. (Geo.) baisasi*

頭毛8(8-C)は頭毛9,10(9-,10-C)のどちらかとほぼ同長．

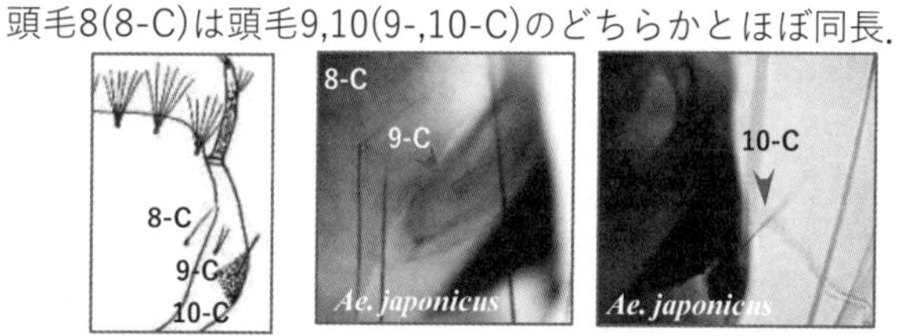

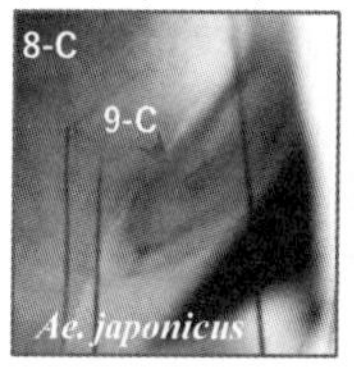

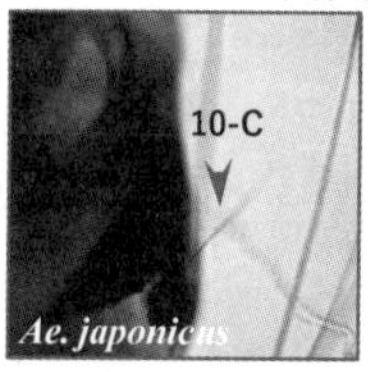

（頭毛の発生位置は種によって多少異なる）

腹面刷毛状毛4(4-X)は15房以上．
(15房のもので頭毛5(5-C)が頭毛7(7-C)より後方に存在するもの)

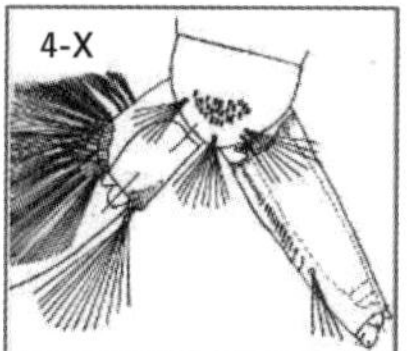

セスジヤブカ亜属 *Ochlerotatus*（一部）
（19ページAへ）

腹面刷毛状毛4(4-X)は15房以下．
(15房のもので頭毛5(5-C)が頭毛7(7-C)より前方に存在するもの)

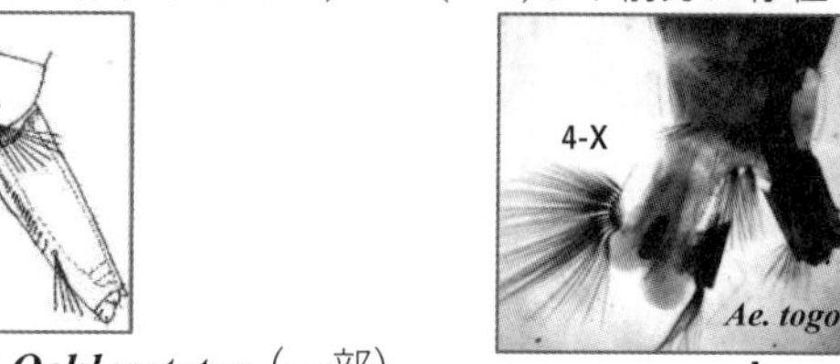

頭毛4～6(4~6-C)のいずれかは頭毛7(7-C)と同じ高さかそれより前方にあり．

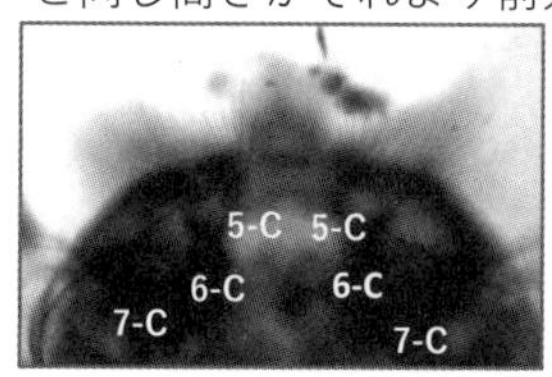

（22ページAへ）

頭毛4～6(4~6-C)の全てが頭毛7(7-C)の後方にあり

ナンヨウヤブカ亜属 *Neomelaniconion*

第Ⅷ腹節毛1,2(1,2-Ⅷ)は同一肥厚板上にあり．

ナンヨウヤブカ *Ae. (Neo.) lineatopennis*

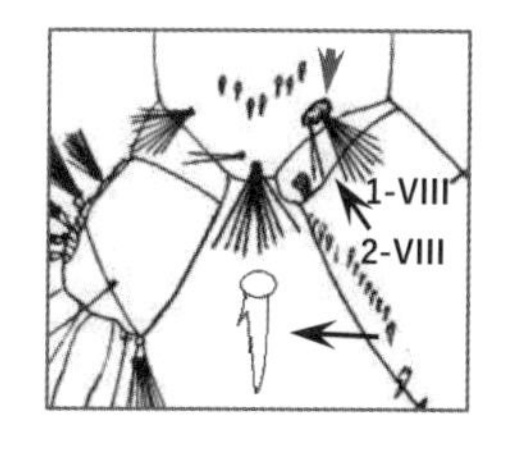

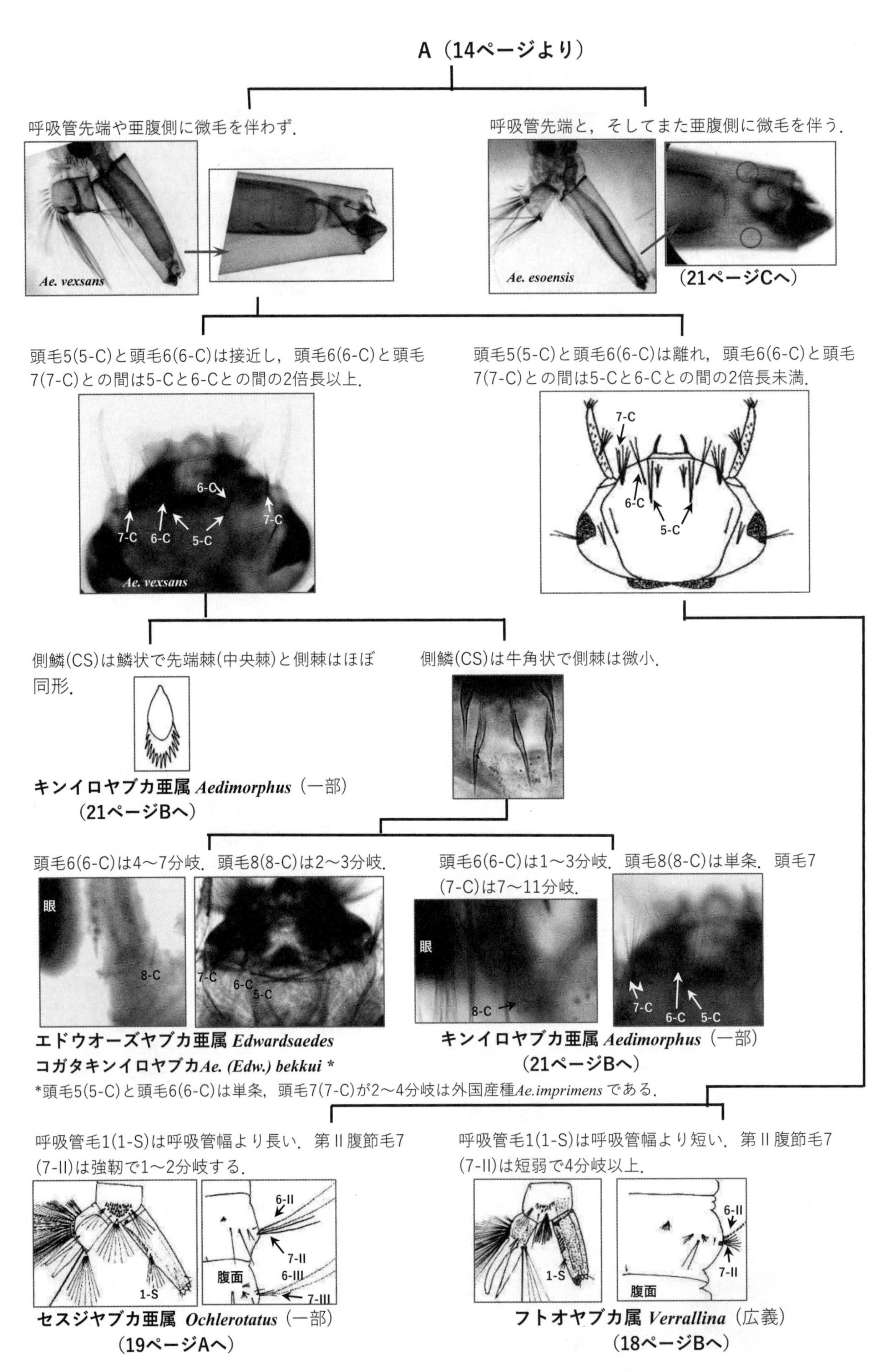
A（14ページより）
呼吸管先端や亜腹側に微毛を伴わず.
Ae. vexsans
呼吸管先端と，そしてまた亜腹側に微毛を伴う.
Ae. esoensis
（21ページCへ）
頭毛5(5-C)と頭毛6(6-C)は接近し，頭毛6(6-C)と頭毛7(7-C)との間は5-Cと6-Cとの間の2倍長以上.
6-C
7-C
7-C
6-C
5-C
Ae. vexsans
頭毛5(5-C)と頭毛6(6-C)は離れ，頭毛6(6-C)と頭毛7(7-C)との間は5-Cと6-Cとの間の2倍長未満.
7-C
6-C
5-C
側鱗(CS)は鱗状で先端棘(中央棘)と側棘はほぼ同形.
キンイロヤブカ亜属 Aedimorphus（一部）
（21ページBへ）
側鱗(CS)は牛角状で側棘は微小.
頭毛6(6-C)は4～7分岐．頭毛8(8-C)は2～3分岐.
眼
8-C
7-C
6-C
5-C
エドウオーズヤブカ亜属 Edwardsaedes
コガタキンイロヤブカAe. (Edw.) bekkui *
頭毛6(6-C)は1～3分岐．頭毛8(8-C)は単条．頭毛7(7-C)は7～11分岐.
眼
8-C
7-C
6-C
5-C
キンイロヤブカ亜属 Aedimorphus（一部）
（21ページBへ）
*頭毛5(5-C)と頭毛6(6-C)は単条，頭毛7(7-C)が2～4分岐は外国産種Ae.imprimensである.
呼吸管毛1(1-S)は呼吸管幅より長い．第Ⅱ腹節毛7(7-II)は強靭で1～2分岐する.
1-S
6-II
7-II
6-III
腹面
7-III
セスジヤブカ亜属 Ochlerotatus（一部）
（19ページAへ）
呼吸管毛1(1-S)は呼吸管幅より短い．第Ⅱ腹節毛7(7-II)は短弱で4分岐以上.
1-S
6-II
7-II
腹面
フトオヤブカ属 Verrallina（広義）
（18ページBへ）

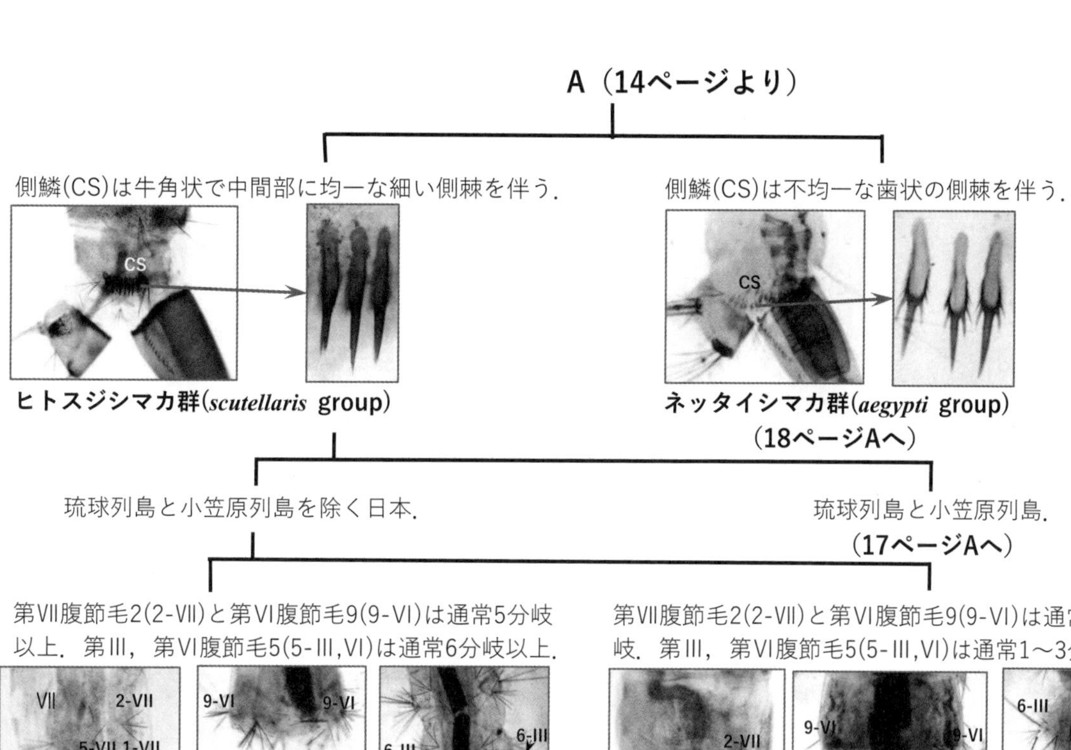

ヒトスジシマカ群(*scutellaris* group)

ネッタイシマカ群(*aegypti* group)
（18ページAへ）

琉球列島と小笠原列島を除く日本.

琉球列島と小笠原列島.
（17ページAへ）

第VII腹節毛2(2-VII)と第VI腹節毛9(9-VI)は通常5分岐以上．第III，第VI腹節毛5(5-III,VI)は通常6分岐以上.

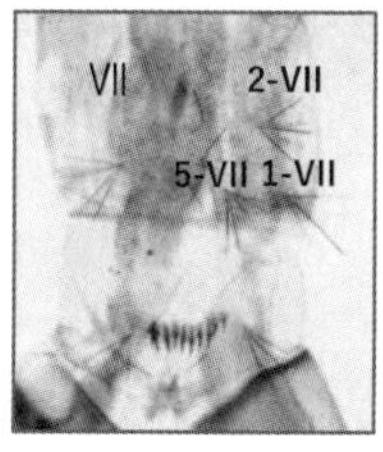

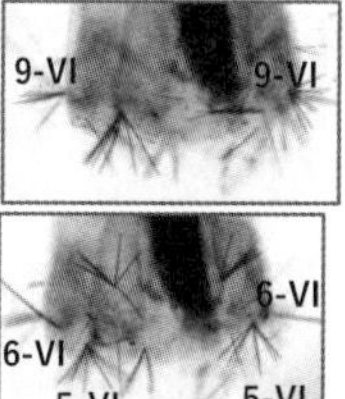

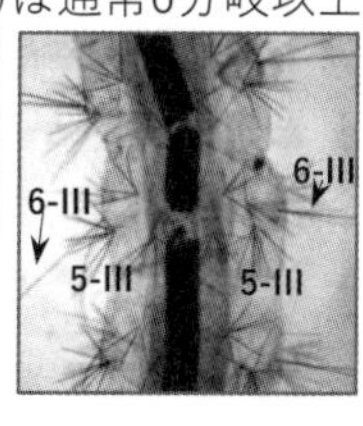

第VII腹節毛2(2-VII)と第VI腹節毛9(9-VI)は通常1～2分岐．第III，第VI腹節毛5(5-III,VI)は通常1～3分岐.

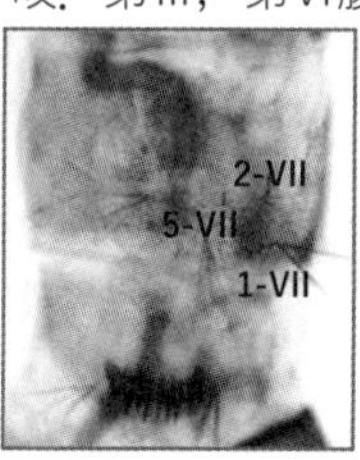

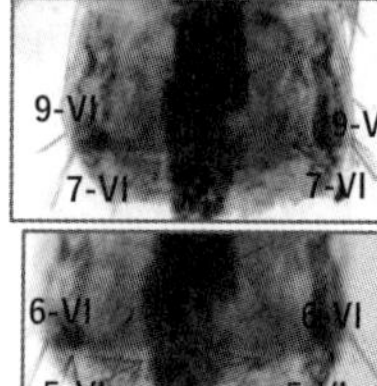

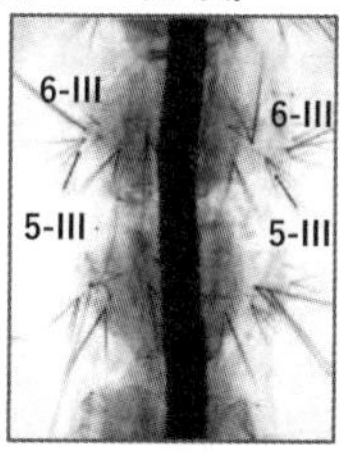

前胸毛4(4-P)は1～2分岐．前胸毛14(14-P)は2～3分岐．鞍板は通常完全な環をなす．(若齢は不完全)

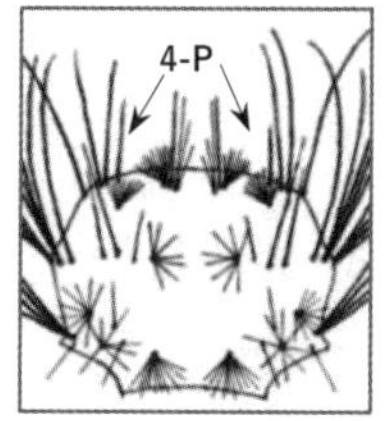

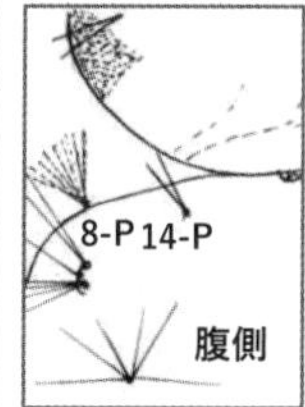

ミズジシマカ *Ae. (Stg.) galloisi*

前胸毛4(4-P)は3～18分岐．前胸毛14(14-P)は4～16分岐．鞍板の環は不完全.

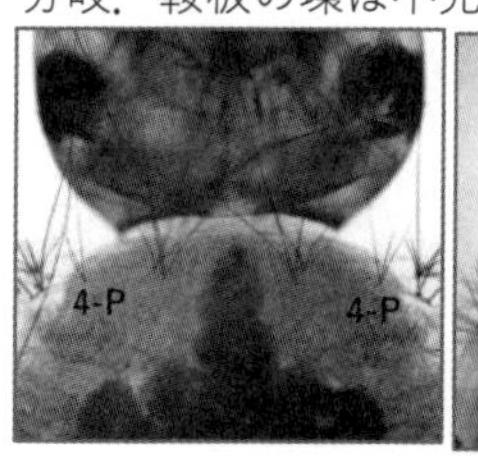

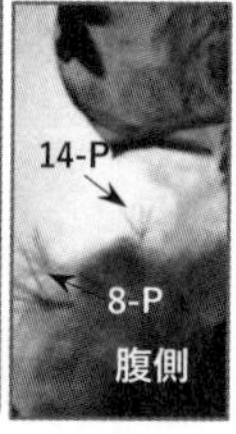

ヤマダシマカ *Ae. (Stg.) flavopictus*（一部）
（17ページBへ）

頭毛6(6-C)は通常単条．稀に側鱗(CS)のいくつかは中央棘先端が裂ける．また中央棘の長さは通常基部より短い．中央棘の微細な側棘は中央棘の基部側1/2を越えて生えるものが多い.

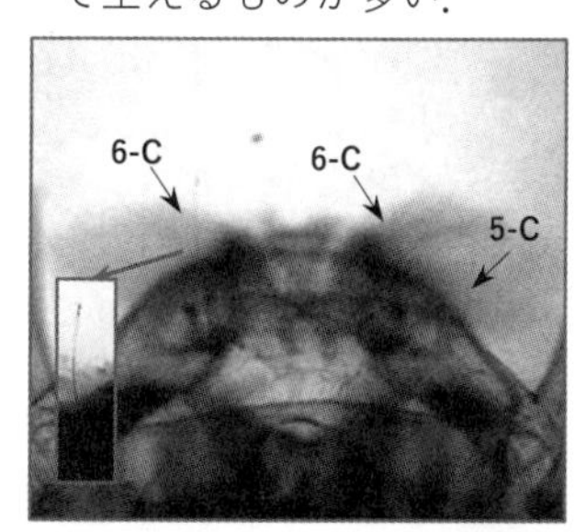

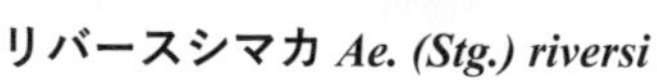
リバースシマカ *Ae. (Stg.) riversi*

頭毛6(6-C)は通常2(1～3)分岐．通常側鱗(CS)の中央棘先端は裂けず(稀に1,2鱗裂けるものあり)，また中央棘は通常基部とほぼ同長か長い．中央棘の微細な側棘は通常中央棘の基部側1/2を越えて生えない.

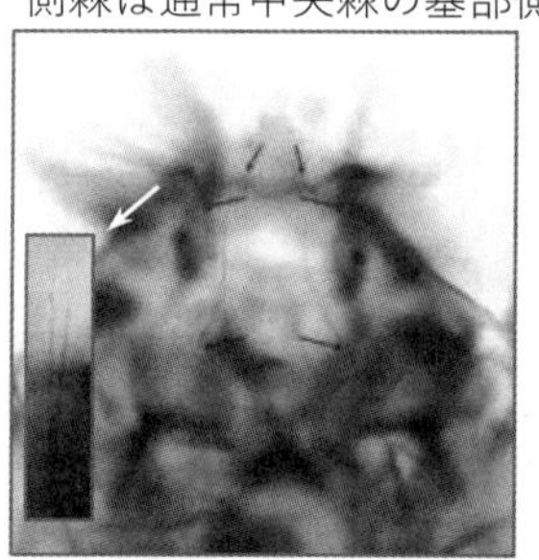
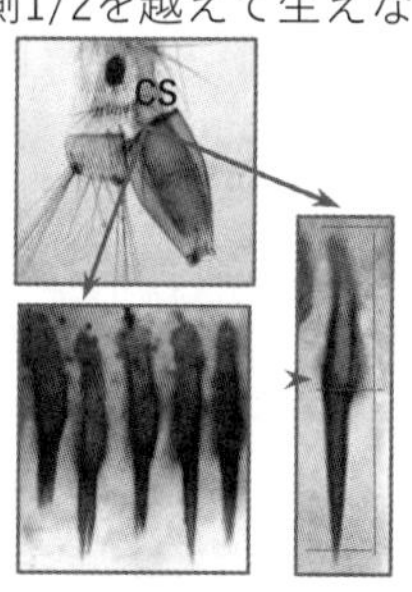

ヒトスジシマカ *Ae. (Stg.) albopictus*
(6-Cは2本が密着していたり2本になりかけのものは1本に見える)

- 16 -

A（16ページより）　B（16ページより）

頭毛6(6-C)は通常単条．2分岐のもの通常側鱗(CS)の中央棘は基部より短いか先が裂けるか，または前胸毛8(8-P)が3～4分岐する．

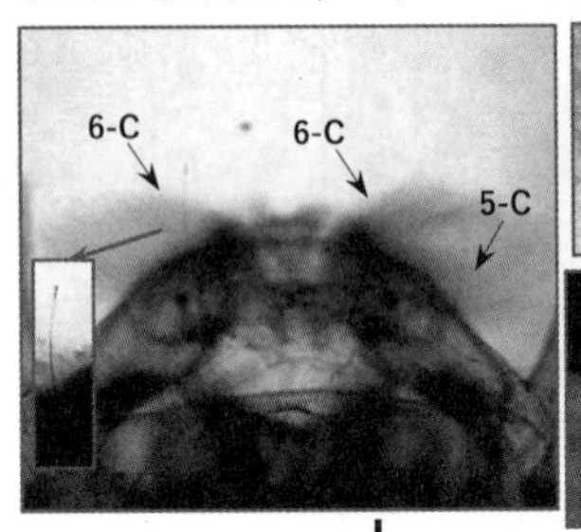

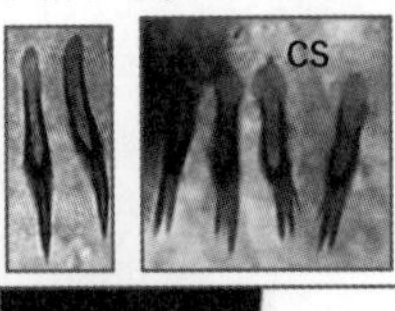

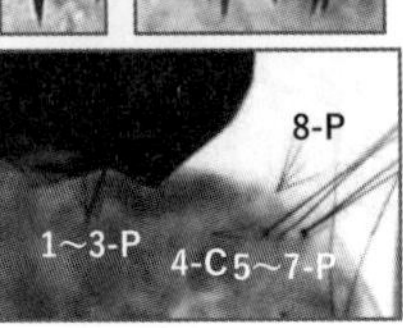

頭毛6(6-C)は通常2分岐．単条のもの通常側鱗(CS)の中央棘は基部とほとんど同じか長く，先が裂けないか，または前胸毛8(8-P)が5分岐以上．

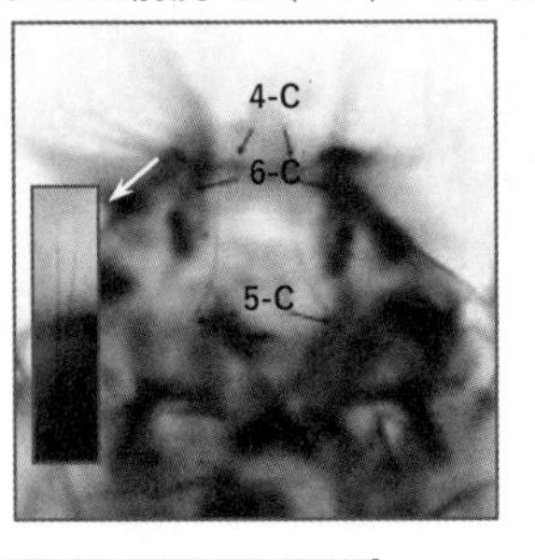

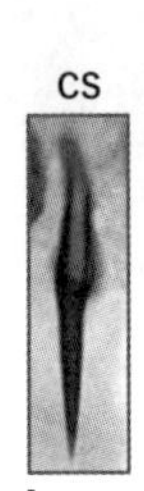

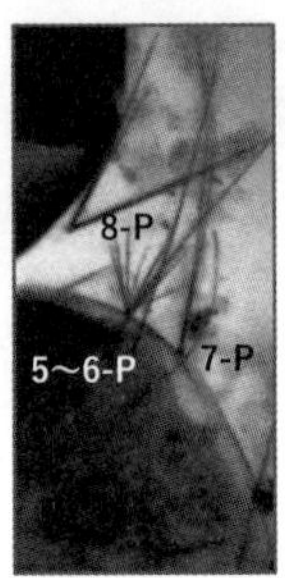

第Ⅷ腹節毛5(5-Ⅷ)は1～2分岐．前胸毛5(5-P)は1～2分岐(片側2分岐多し)．前胸毛7(7-P)は2～3分岐(片側3分岐あり)．多くの側鱗(CS)の中央棘が基部と同長か長く先端は裂けない．

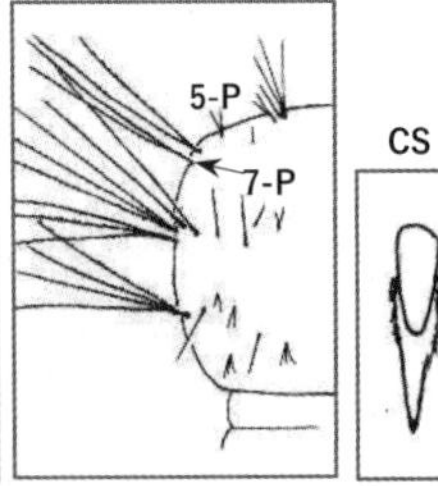

ダイトウシマカ ***Ae. (Stg.) daitensis***

第Ⅷ腹節毛5(5-Ⅷ)は3～6分岐．前胸毛5(5-P)は単条．前胸毛7(7-P)は2分岐．通常側鱗(CS)の多くは中央棘の長さが基部より短いか先端が裂ける．

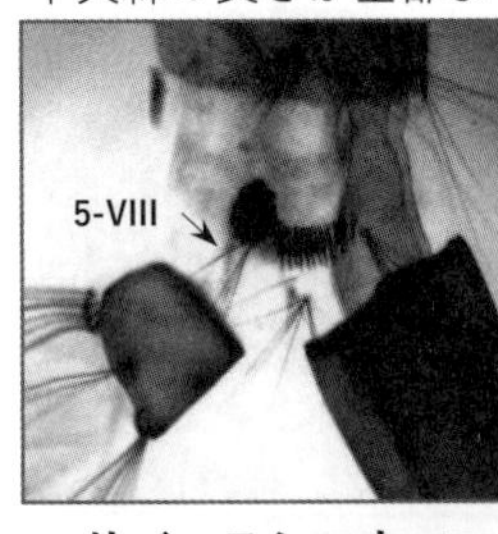

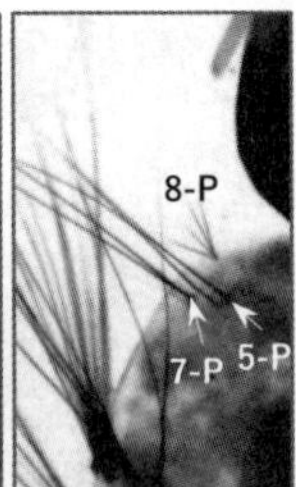

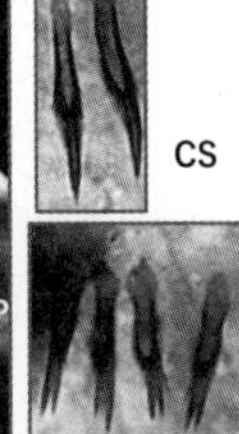

リバースシマカ ***Ae. (Stg.) riversi***

頭毛10(10-C)は通常単条．通常腹側尾葉は背側尾葉と同長．

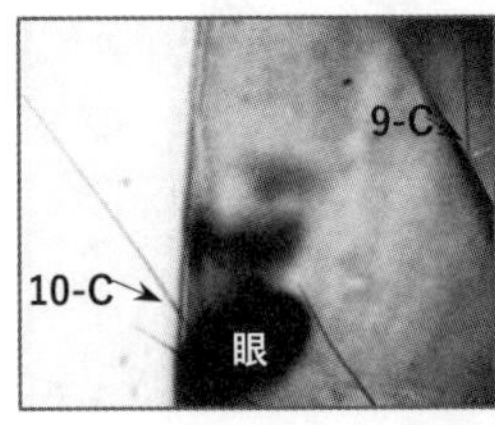

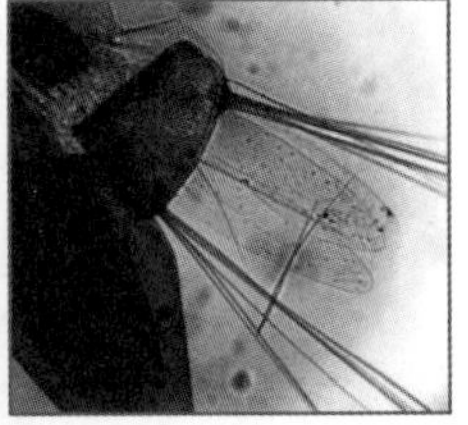

ヒトスジシマカ ***Ae. (Stg.) albopicus***

頭毛10(10-C)は通常2分岐．通常腹側尾葉は背側尾葉より短い．

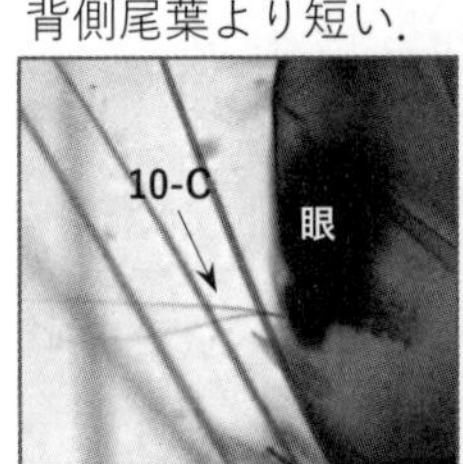

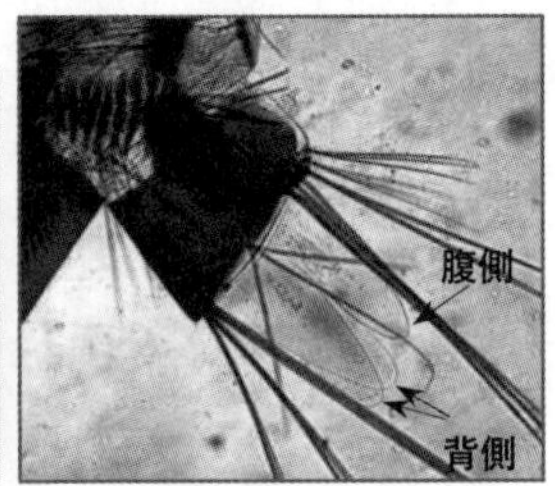

ヤマダシマカ群（*flavopictus* groop）

尾葉は短く通常鞍板幅に達しない．腹節毛はあまり毛深くは見えない．

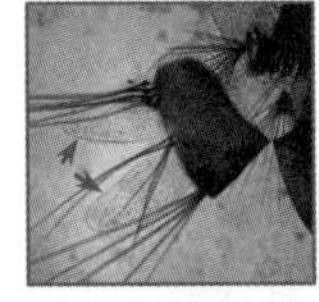
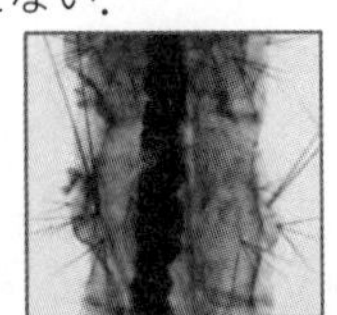

沖縄島，奄美大島に分布．

ダウンスシマカ ***Ae. (Stg.) flavopictus downsi***

尾葉はやや長く通常鞍板幅と等しいか長い．腹節毛は毛深く見えるかまたは細い．

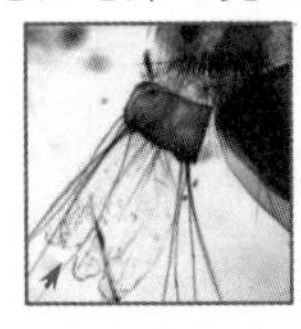
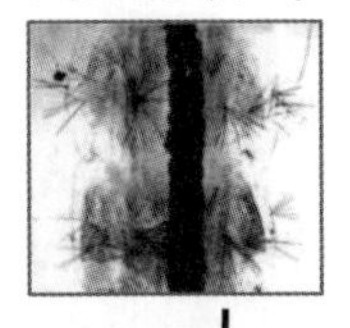

腹節毛は通常剛毛で放射状に広がる．

九州以北にに分布

ヤマダシマカ ***Ae. (Stg.) flavopictus flavopictus***

腹節毛はやや細いか長い．

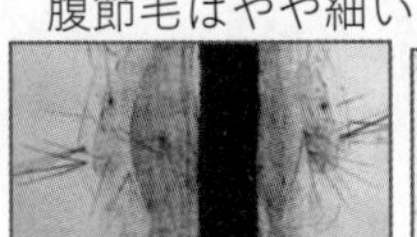

沖縄八重山群島に分布．

ミヤラシマカ ***Ae. (Stg.) flavopictus miyarai***

A（16ページより）
ネッタイシマカ群(*aegypti* groop)

B（15ページより）
フトオヤブカ属 *Verrallina*（広義）

前胸毛8(8-P)は3～4分岐，14(14-P)は2～4分岐．第VI，VII腹節毛2(2-VI, -VII)は単条．

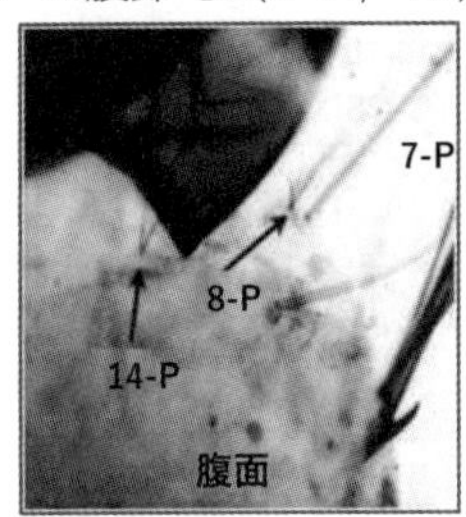

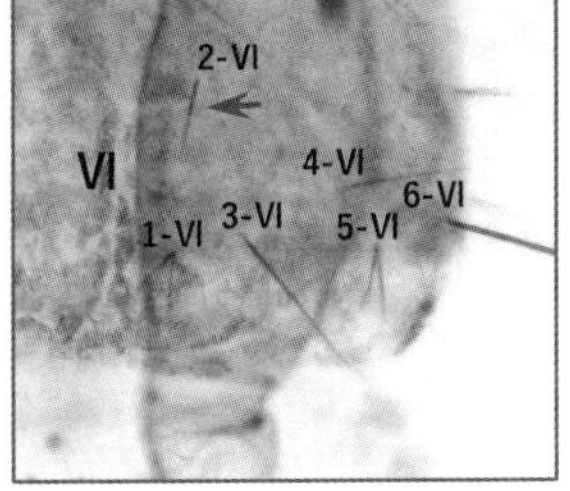

ネッタイシマカ *Ae. (Stg.) aegypti*

前胸毛8(8-P)は8～11分岐，14(14-P)は6～15分岐．第VI，VII腹節毛2(2-Vil, -VII)は3～7分岐．

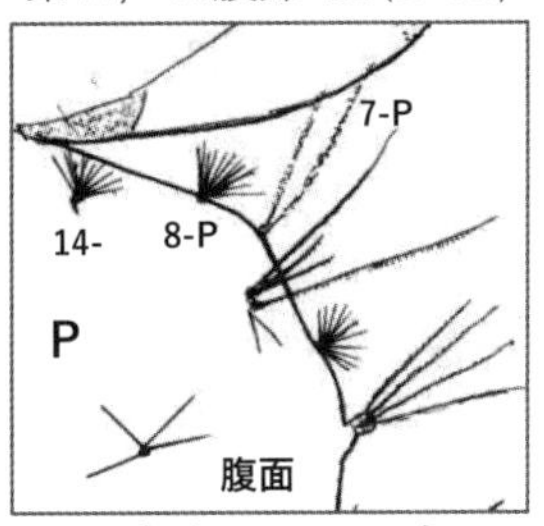

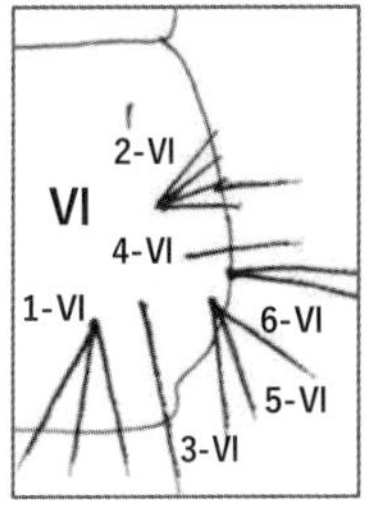

タカハシシマカ *Ae. (Stg.) wadai*

クロフトオヤブカ亜属 *Verrallina*，アカフトオヤブカ亜属 *Neomacleana*，コガタフトオヤブカ亜属 *Harbachius*

頭毛5,6(5,6-C)は1～2分岐．触角毛1(1-A)は単条か2分岐．背面刷毛状毛2(2-X)は3～4分岐．

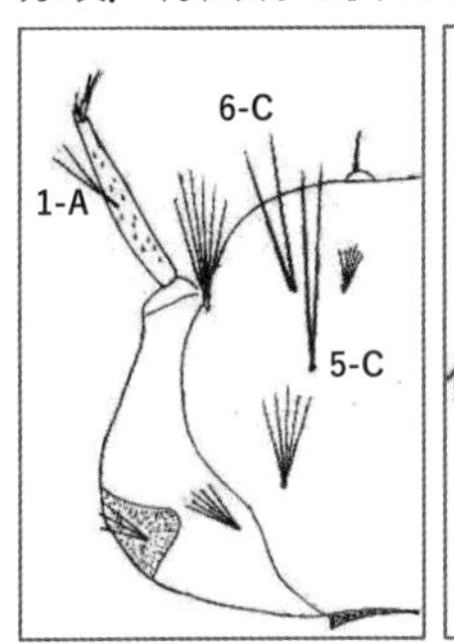

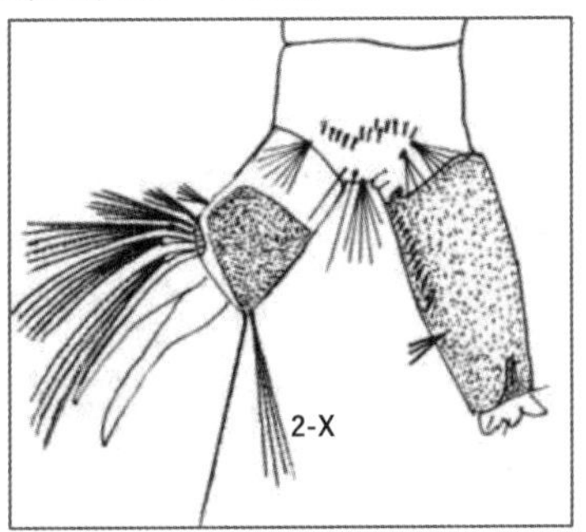

コガタフトオヤブカ亜属 *Harbachius*
コガタフトオヤブカ *Ve. (Har.) nobukonis*

頭毛5,6(5,6-C)は3分岐以上．触角毛1(1-A)は3分岐以上．背面刷毛状毛2(2-X)は5～7分岐．

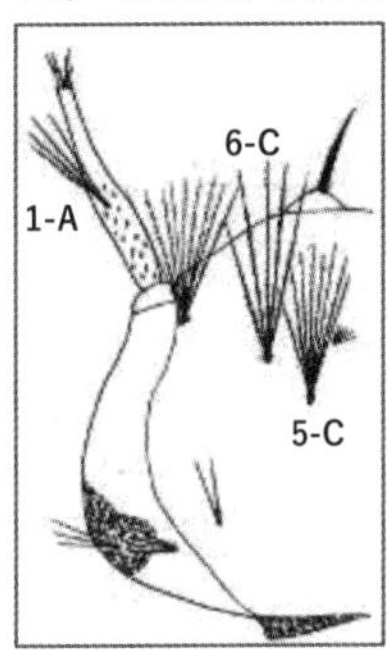

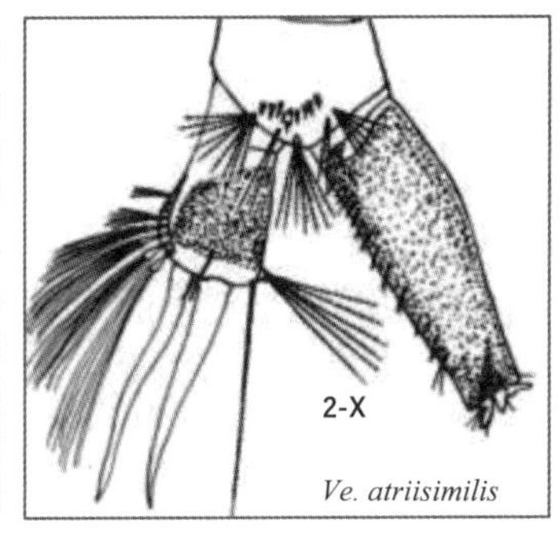

頭毛5(5-C)は5～8分岐．頭毛6(6-C)は5～7分岐．頭毛7(7-C)は11～16分岐．呼吸管比は3.0～3.5．呼吸管棘(PT)は13～16個で呼吸管基部から約3/4位置まで存在．

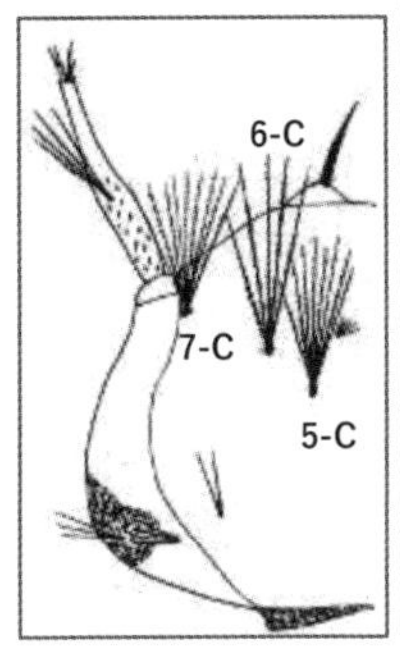

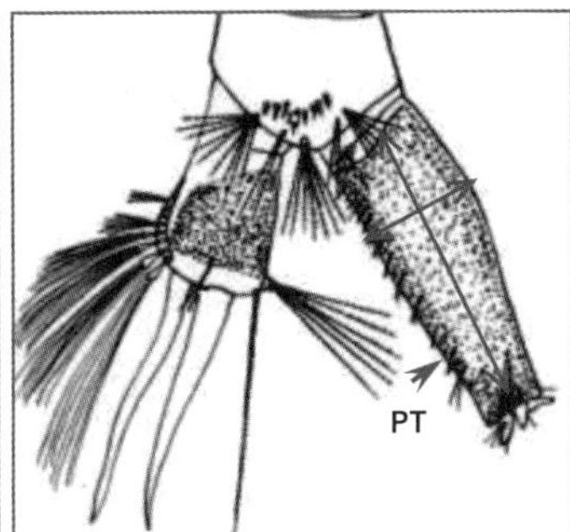

アカフトオヤブカ亜属 *Neomacleana*
アカフトオヤブカ *Ve. (Neo.) atriisimilis*
（沖縄県西表島から採集報告あり）

頭毛5(5-C)は3～4分岐．頭毛6(6-C)は3～4分岐．頭毛7(7-C)は5～7分岐．呼吸管比は2.0～3.0．呼吸管棘(PT)は10～14個で呼吸管基部から約1/2位置まで存在．

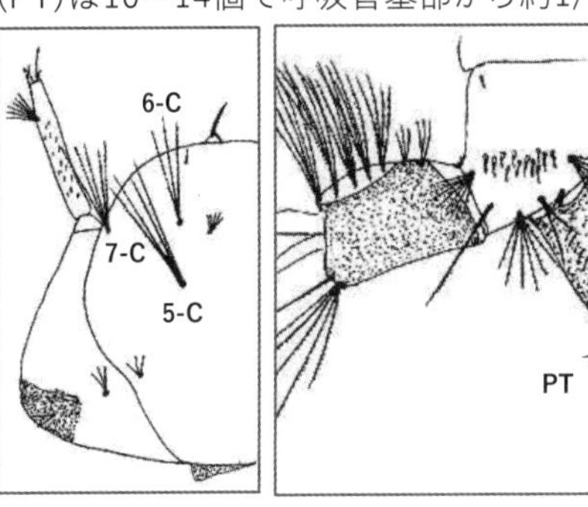

クロフトオヤブカ亜属 *Verrallina*
クロフトオヤブカ *Ve. (Ver.) iriomotensis*
（沖縄県西表島から採集報告あり）

- 18 -

A（14,15ページより）

セスジヤブカ亜属 *Ochlerotatus*

側鱗(CS)の先端棘(中央棘)は他の側棘とほとんど同形で，先端棘近くの側棘の長さは先端棘の1/2長以上.

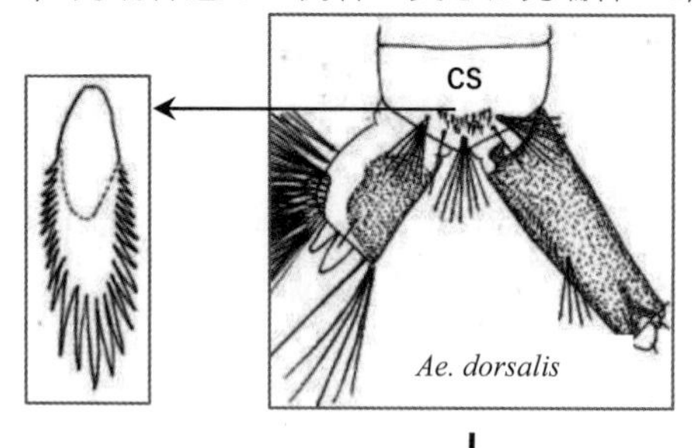

側鱗(CS)の先端棘(中央棘)は他の側棘に比べ強靭で大きく(先端が2分岐するものあり),他の側棘は先端棘の1/2長に満たない.

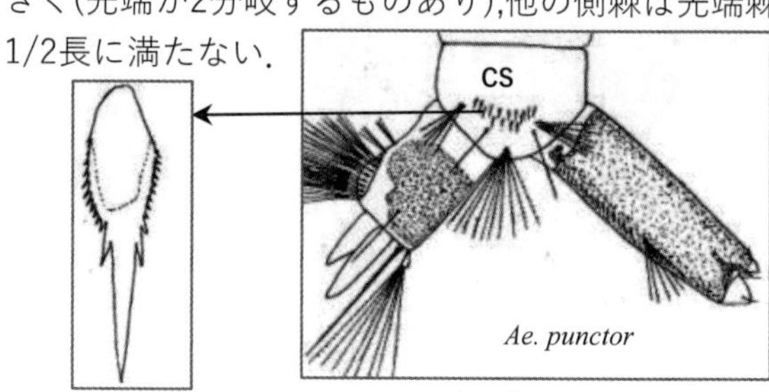

前胸毛1(1-P)は2分岐で強靭(北米のもの3分岐あり).第Ⅲ～第Ⅴ腹節毛13(13-Ⅲ～13-Ⅴ)は強靭.側鱗(CS)は28～81個で通常は40個以上.

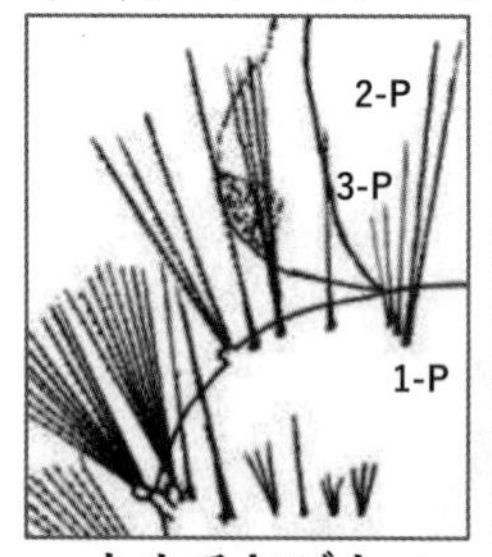

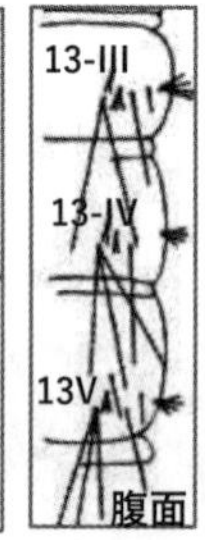

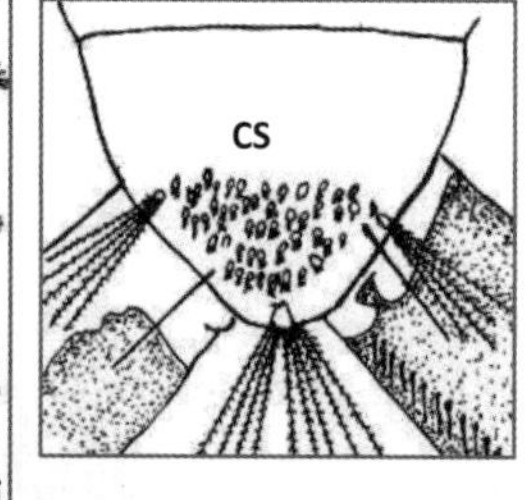

トカチヤブカ *Ae. (Och.) communis*

前胸毛1(1-P)は単条で弱々しい.第Ⅲ～第Ⅴ腹節毛13(13-Ⅲ～13-Ⅴ)は弱々しい.側鱗(CS)は15～33個.

(北米のセスジヤブカは1-Pは強靭)

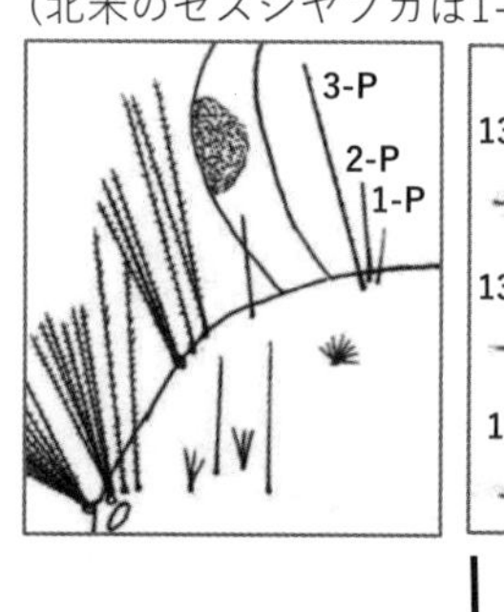

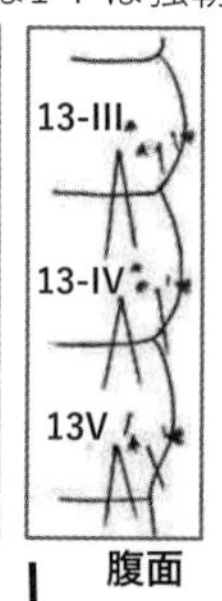

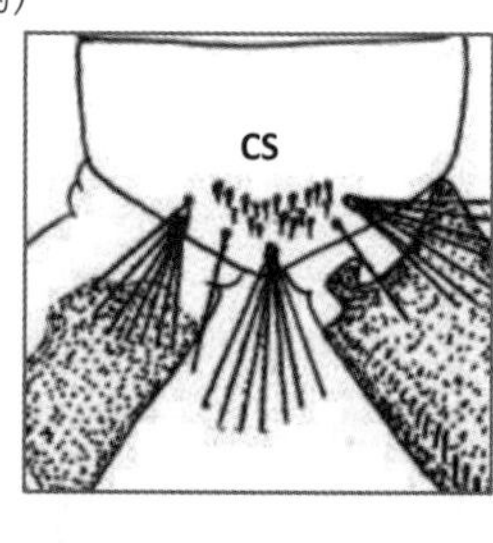

腹面

触角毛1(1-A)は2～3分岐.呼吸管棘(PT)は7～12個で先端側の棘は基部側32%～42%の位置に存在する.

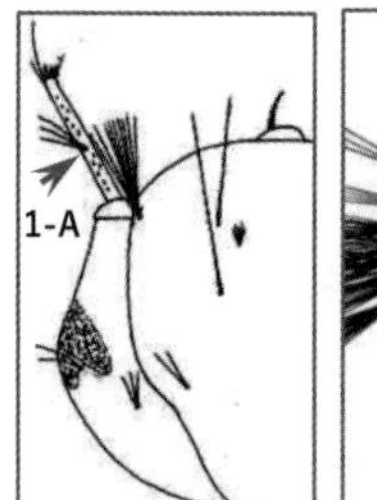

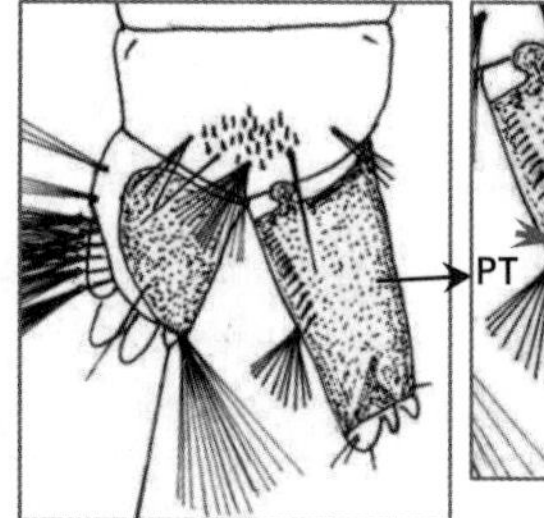

ハマベヤブカ *Ae. (Och.) vigilax*

触角毛1(1-A)は5～12分岐.呼吸管棘(PT)は16～28個で先端側の棘は基部側45%～51%の位置に存在する.

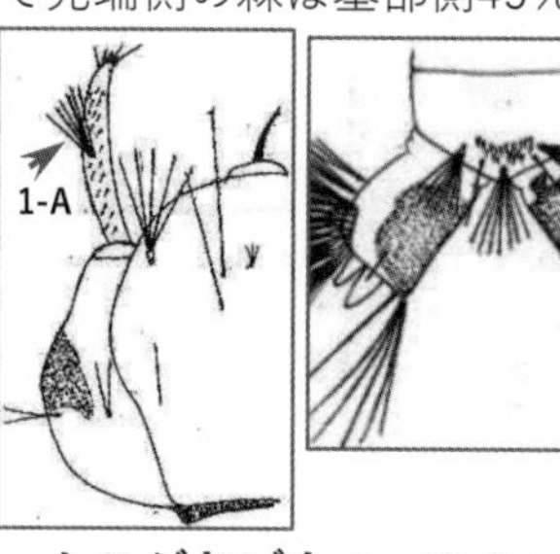

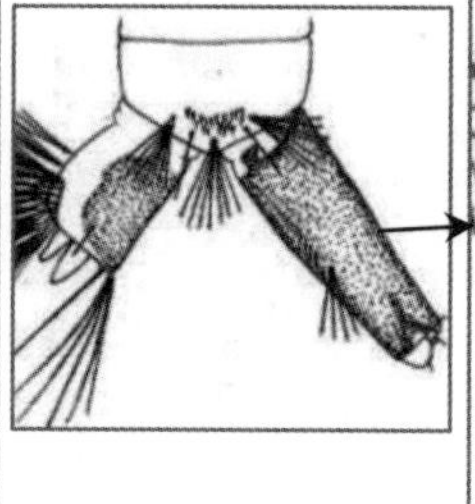

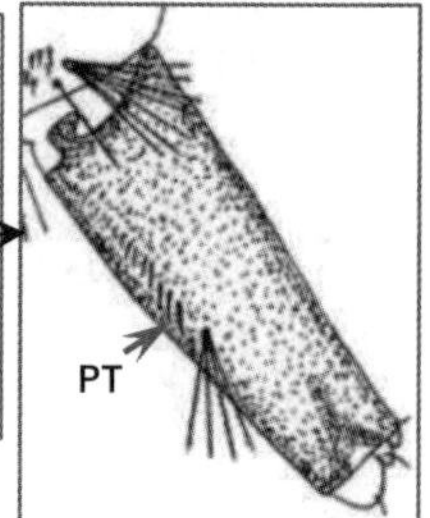

セスジヤブカ*Ae. (Och.) dorsalis*

呼吸管棘(PT)はほぼ等間隔に生じる.

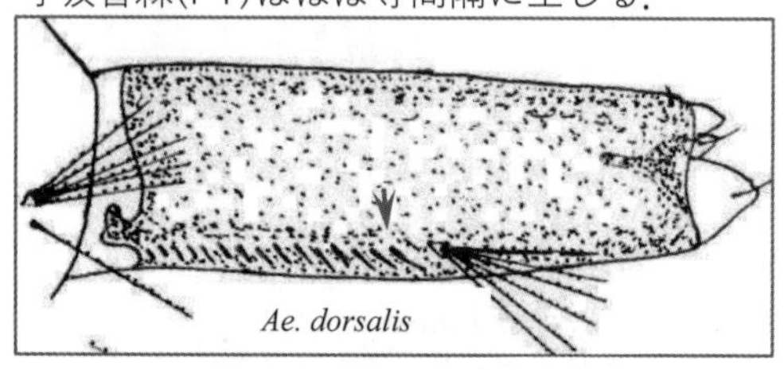

(20ページAへ)

呼吸管棘(PT)先端棘1ないし数個は明確に離れて生じる.

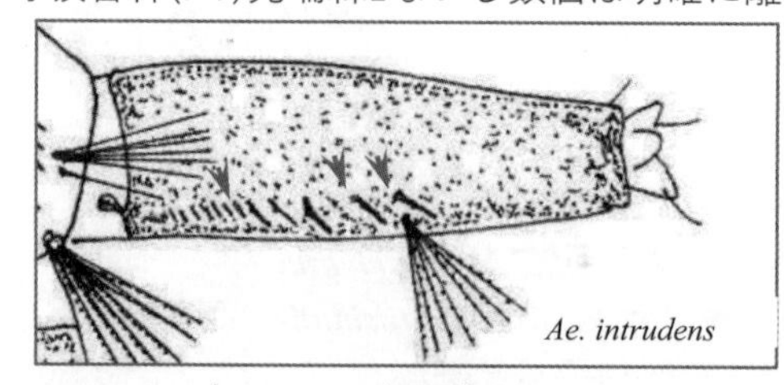

(21ページAへ)

A（19ページより）

前胸毛2,3(2-,3-P)は明確に前胸毛1(1-P)より細くて短い．鞍板毛1(1-X)は細く鞍板長より短い．

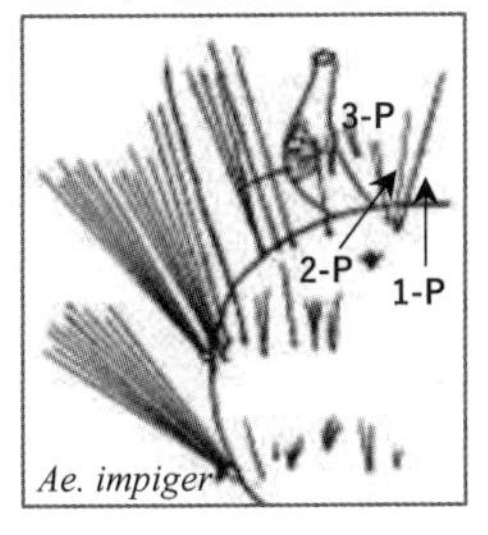

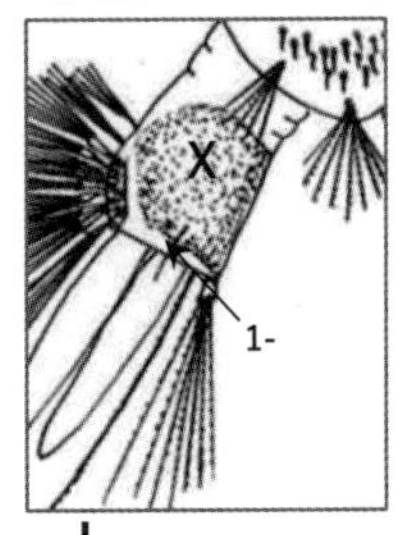

前胸毛2,3(2-,3-P)のどちらか一方あるいは双方が前胸毛1(1-P)とほぼ同じ太さと長さ．鞍版毛1(1-X)は鞍板長と同じか長い．

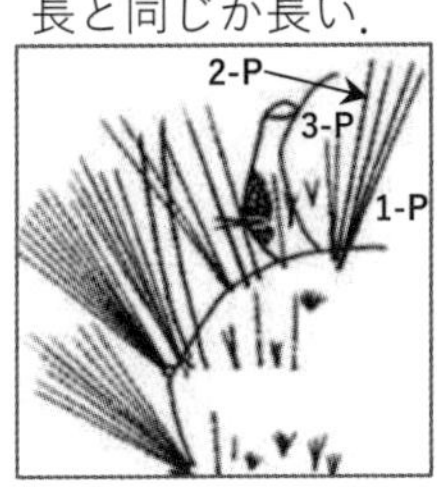

頭毛5(5-C)は単条で幾分太い．第Ⅰ腹節毛12(12-I)は存在しない．側鱗(CS)は9〜17個．

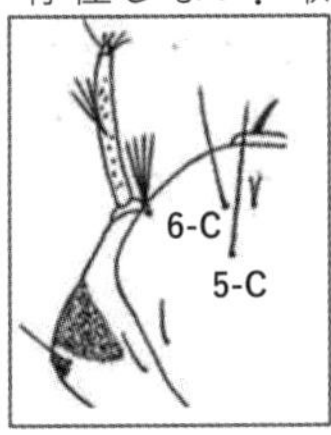

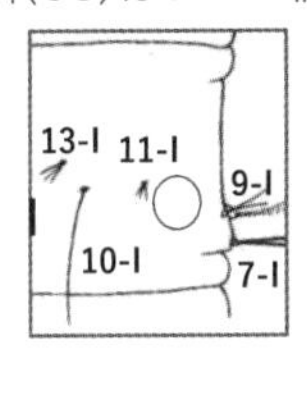

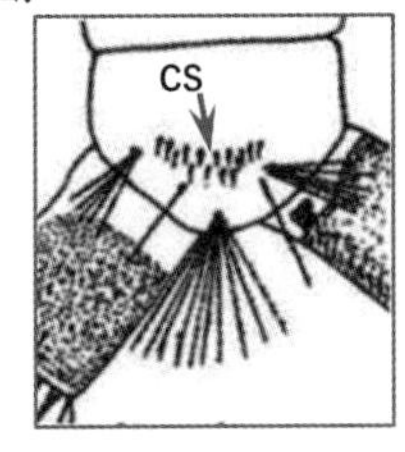

ダイセツヤブカ *Ae. (Och.) impiger daisetsuzanus*

頭毛5(5-C)は2〜4分岐で細い．第Ⅰ腹節毛12(12-I)は存在する．側鱗(CS)は18〜25個．

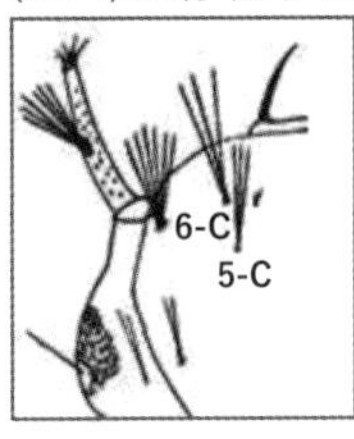

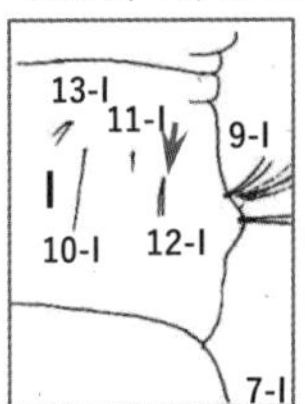

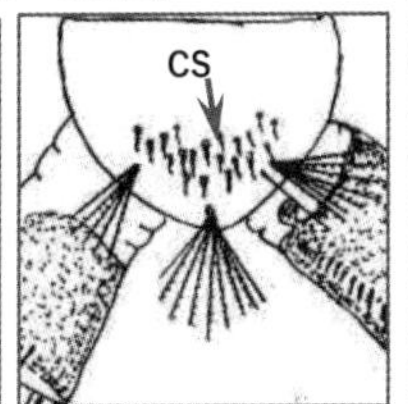

カラフトヤブカ *Ae. (Och.) sticticus*

中胸毛1(1-M)は中胸毛3(3-M)より太く明らかに長い．

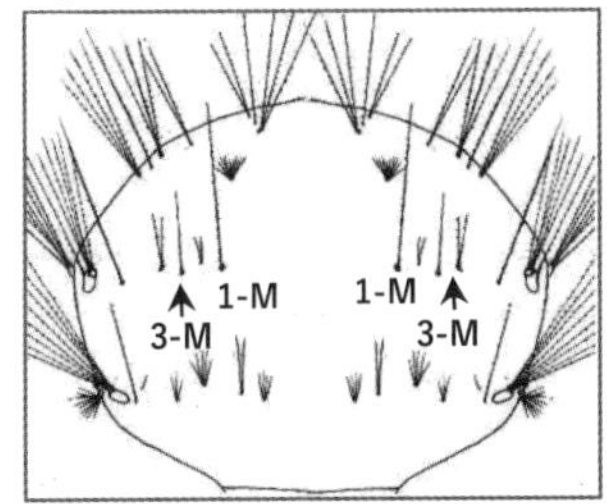

ハクサンヤブカ *Ae. (Och.) hakusanensis*

中胸毛1(1-M)は中胸毛3(3-M)より短く弱々しい．

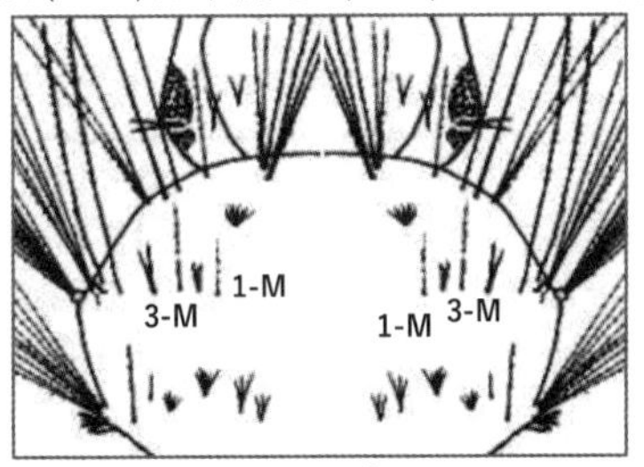

第Ⅳ,Ⅴ腹節毛6(6-Ⅳ,-Ⅴ)は単条．第Ⅲ腹節毛６(６-Ⅲ)はほとんど常に単条，非常にまれに２分岐．側鱗(CS)は6〜12個，通常は10個以上．

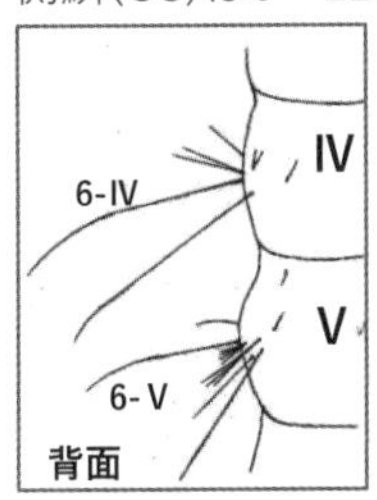

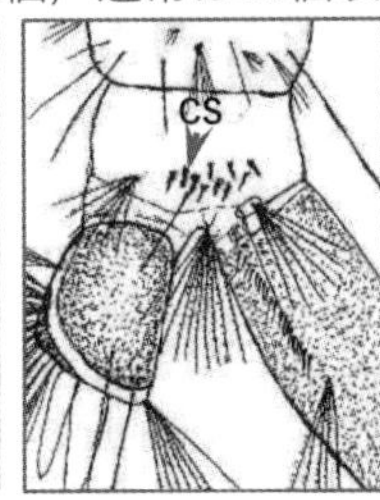

チシマヤブカ *Ae. (Och.) punctor*

第Ⅲ〜Ⅴ腹節毛6(6-Ⅲ〜Ⅴ)はほとんど常に2分岐．

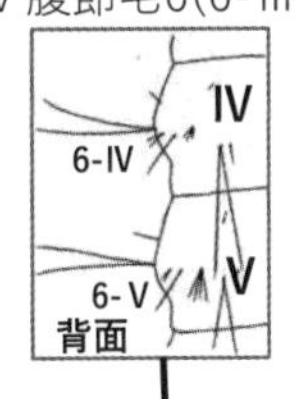

側鱗(CS)は5〜9個．第Ⅴ腹節毛13(13-Ⅴ)はほとんど常に2分岐．前胸毛3(3-P)は通常2〜3分岐．

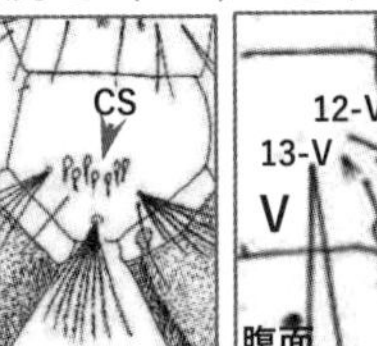

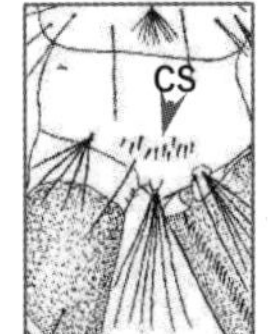

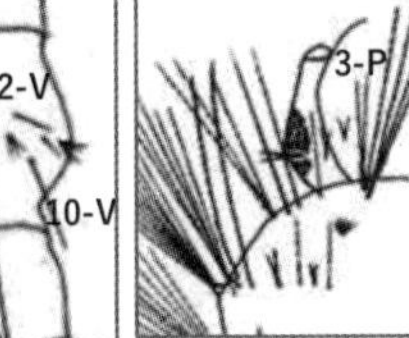

キタヤブカ *Ae. (Och.) hokkaidensis*

側鱗(CS)は8〜16個．第Ⅴ腹節毛13(13-Ⅴ)はほとんど常に単条．前胸毛3(3-P)は通常単条．

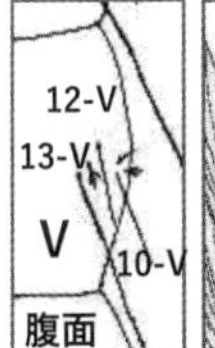

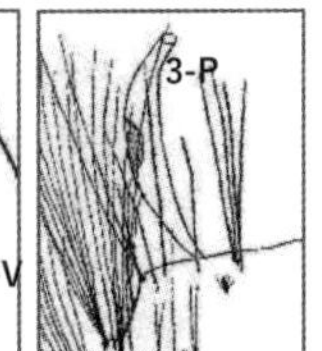

アッケシヤブカ *Ae. (Och.) akkeshiensis*

A（19ページより）

B（15ページより）
キンイロヤブカ亜属
Aedimorphus

C（15ページより）
エゾヤブカ亜属 *Aedes*

側鱗(CS)は28～38個が班をなす．呼吸管比は4.2～4.9．触角毛1(1-A)は6～11分岐．

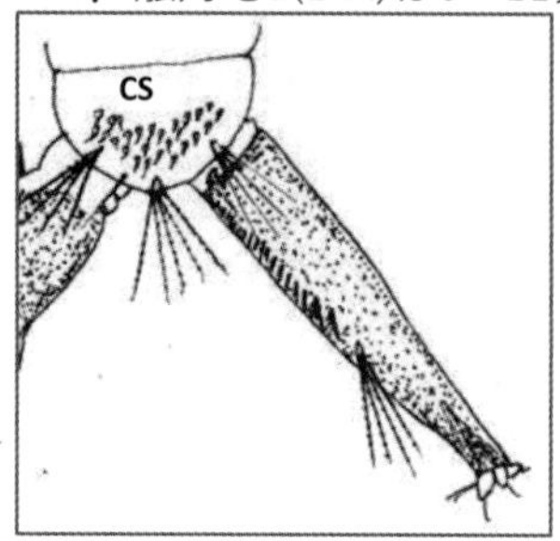

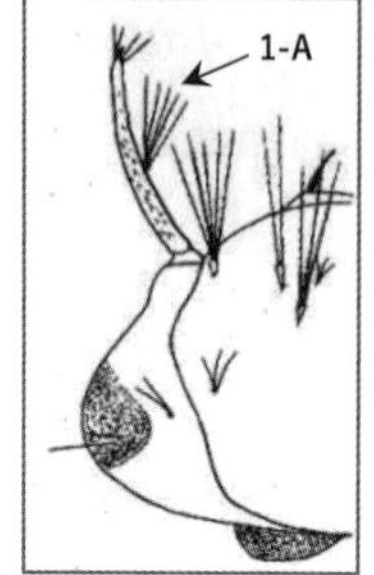

アカンヤブカ *Ae. (Och.) excrucians*

側鱗(CS)は8～16個が1又は2列に不規則に並ぶ．呼吸管比は3.0～4.0．

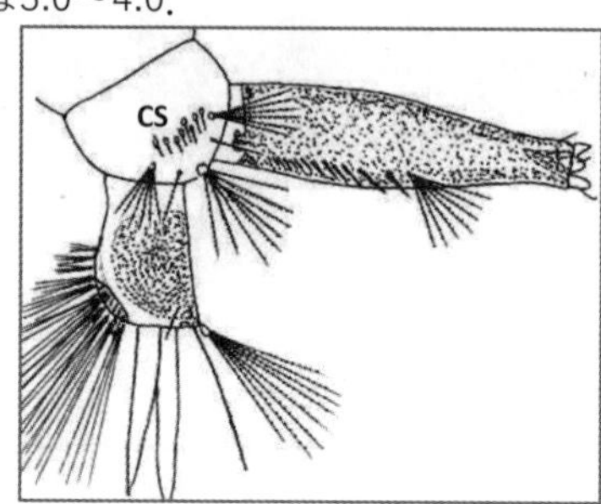

触角は頭長より短い．頭毛5(5-C)は頭毛6(6-C)の垂直かごくわずか正中よりで下方に離れて生じる．

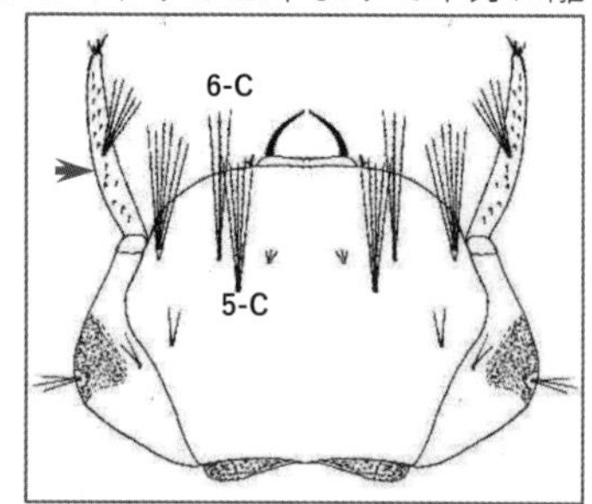

サッポロヤブカ *Ae. (Och.) intrudens*

触角は頭長と同長か長い．頭毛5(5-C)は頭毛6(6-C)の下方で明確に正中との中間部に生じる．

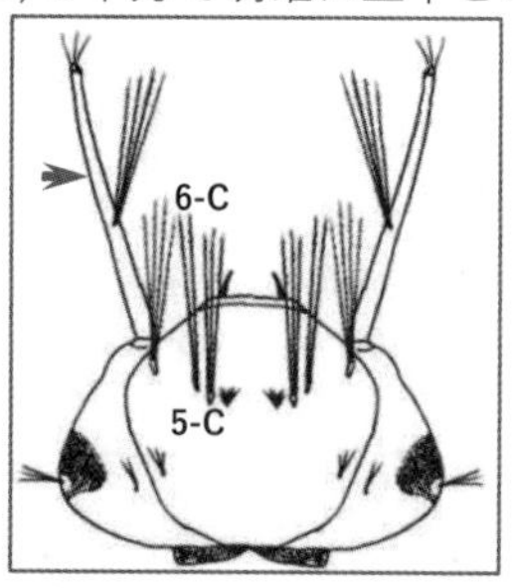

ヒサゴヌマヤブカ *Ae. (Och.) diantaeus*

頭毛5～7(5～7-C)はほぼ斜め一列に並び頭毛5,6(5,6-C)は頭毛4(4-C)のほぼ同じ高さにある．側鱗(CS)は鱗状で先端棘（中央棘）は側棘と同形．第Ⅷ腹節毛3(3-Ⅷ)は第Ⅷ腹節毛1(1-Ⅷ)と第Ⅷ腹節毛5(5-Ⅷ)よい短い．

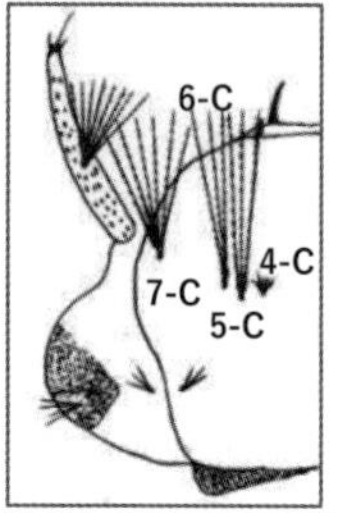

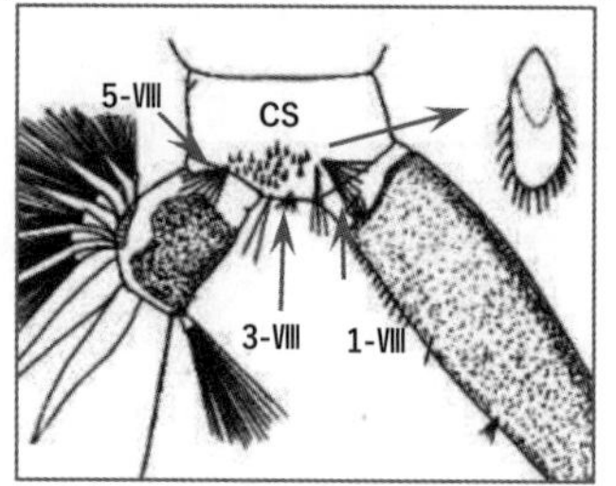

オオムラヤブカ *Ae. (Adm.) alboscutellatus*

頭毛5～7(5～7-C)は一列に並ばないが頭毛4,6,7(4,6,7-C)はほぼ同じ高さにある．側鱗(CS)は牛角状で側棘は微小．第Ⅷ腹節毛3(3-Ⅷ)は第Ⅷ腹節毛1(1-Ⅷ)や第Ⅷ腹節毛5(5-Ⅷ)とほぼ同大．

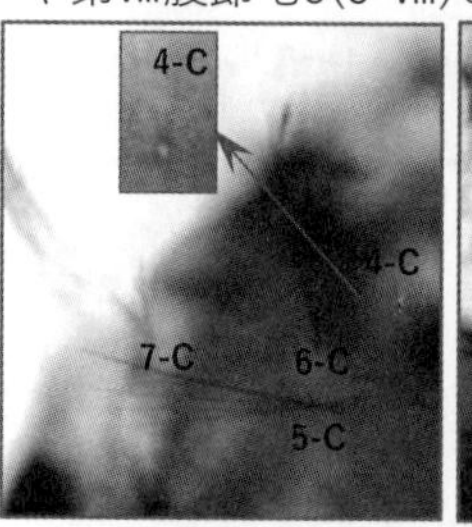

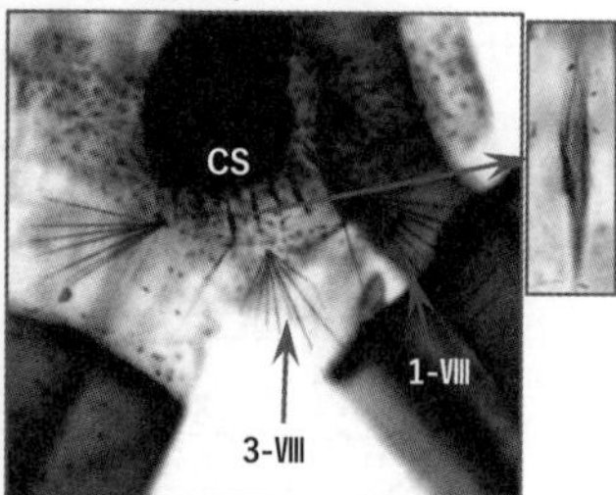

キンイロヤブカ *Ae. (Adm.) vexans nipponii*

後胸毛6(6-T)と第Ⅷ腹節毛4(4-Ⅷ)は2～3分岐．

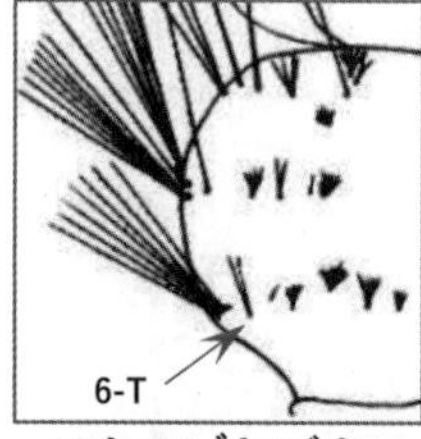

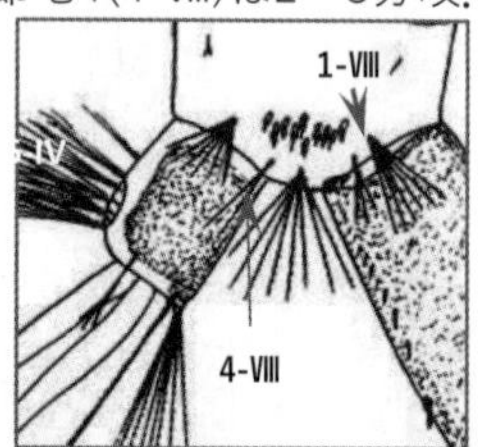

アカエゾヤブカ *Ae. (Aed.) yamadai*

後胸毛6(6-T)と第Ⅷ腹節毛4(4-Ⅷ)は単条．

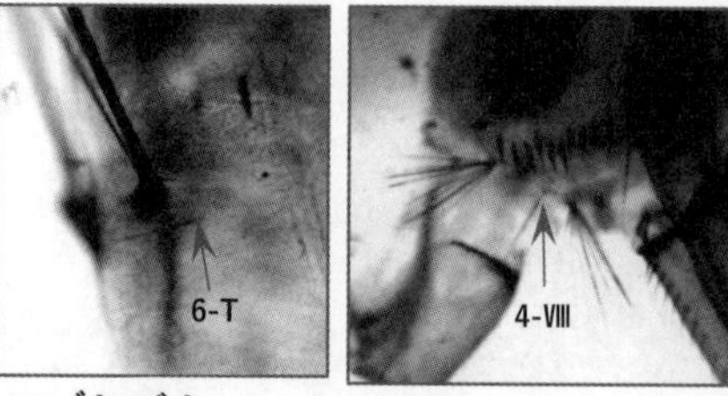

エゾヤブカ *Ae. (Aed.) esoensis**
ホッコクヤブカ *Ae. (Aed.) sasai***

*触角の微棘は棘状．

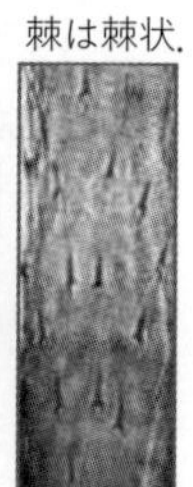

**触角の微棘は微小．

A（14ページより）

トウゴウヤブカ亜属 *Tanakaius*，ヤマトヤブカ亜属 *Hulecoeteomyia*，ハトリヤブカ亜属 *Collessius*，オキナワヤブカ亜属 *Bruceharrisonius*，シロカタヤブカ亜属 *Downsiomyia*，ワタセヤブカ亜属 *Phagomyia*，ケイジョウヤブカ亜属 *Hopkinsius*

呼吸管棘列の数棘は離れて存在し，通常呼吸管毛1(1-S)は呼吸管棘列内に生える．側鱗(CS)は32～93個が班をなし，形状は先端が丸く先端から両側にかけて側棘を有す．

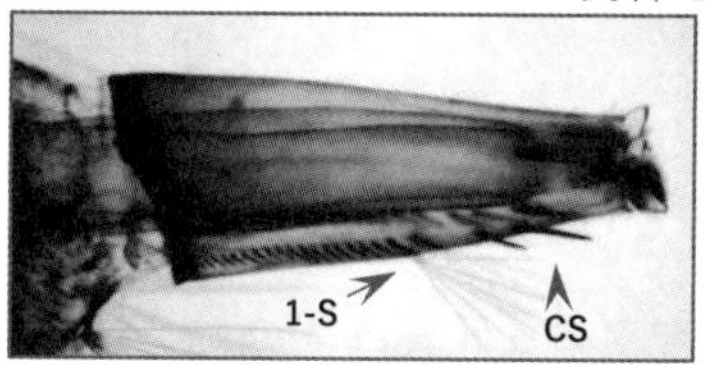

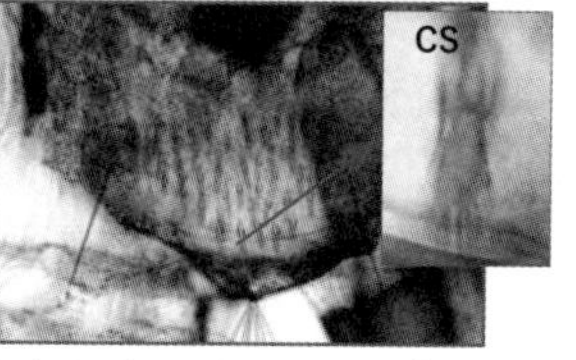

（春先の個体では呼吸管棘が離れて存在しないものがある）

ヤマトヤブカ亜属 *Hulecoeteomyia*

ヤマトヤブカ *Ae. (Hul.) japonicus*

ヤマトヤブカ *Ae. (Hul.) j. japonicus*　：呼吸管比2.67～3.40，前胸毛5(5-P)は2～3分岐，前胸毛7(7-P)は2～4分岐，尾葉は腹背ともほぼ同長．屋久島以北に分布．

アマミヤブカ *Ae. (Hul.) j. amamiensis*：呼吸管比2.58～3.05，前胸毛5(5-P)は単条(稀に2分岐)．前胸毛7(7-P)は単条．尾葉は背側が腹側より長い．奄美大島，徳之島に分布．

サキシマヤブカ *Ae. (Hul.) j. yaeyamensis*：呼吸管比3.41～4.03と3亜種の中で一番長い．石垣島，西表島に分布．

呼吸管棘列の数棘はほぼ均等に並ぶ，呼吸管毛1(1-S)は通常呼吸管棘(PT)と呼吸管先端との間にある．

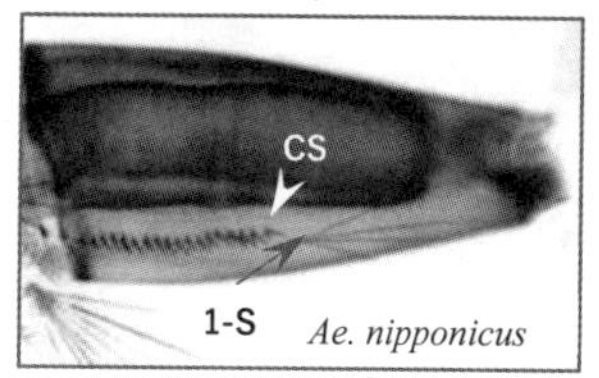

側鱗はほぼしゃもじ状で．先端棘(中央棘)と側棘はほぼ同形．頭毛4(4-C)は非常に小さい．

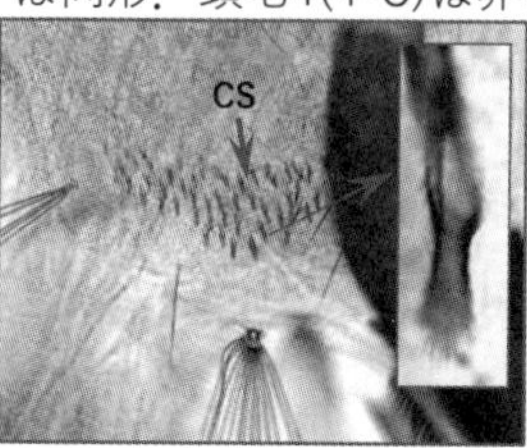

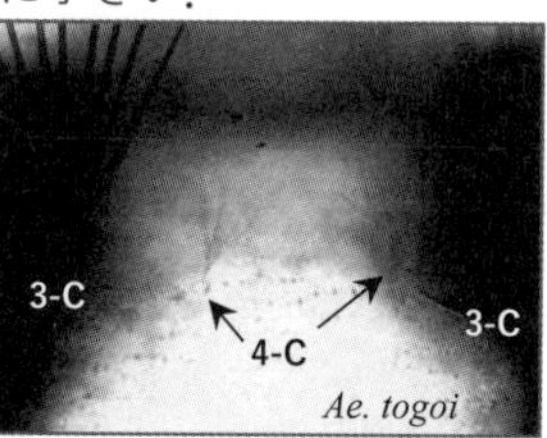

側鱗は牛角状で．先端棘(中央棘)は非常に大きい．頭毛4(4-C)はよく発達する．

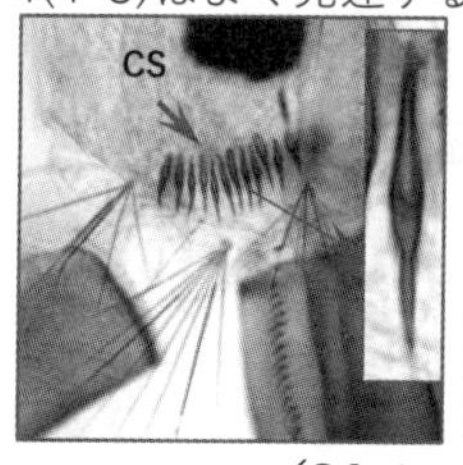

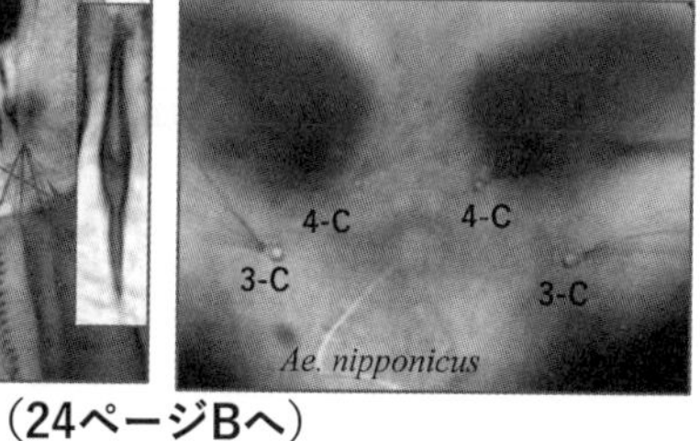

（24ページBへ）

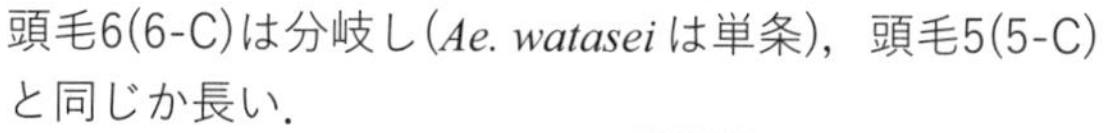

頭毛6(6-C)は分岐し(*Ae. watasei* は単条)，頭毛5(5-C)と同じか長い．

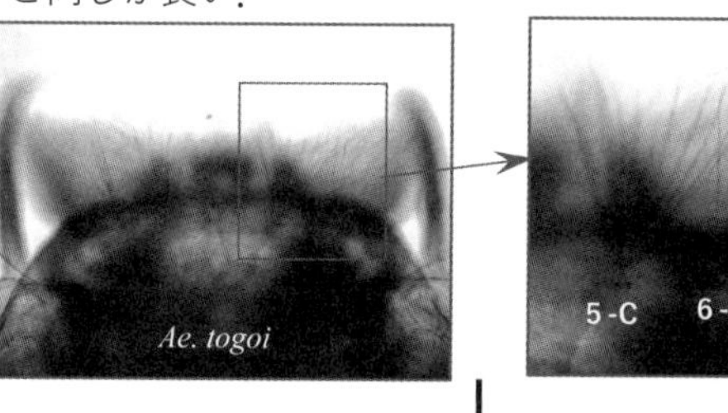

頭毛6(6-C)は単条で，頭毛5(5-C)より遥かに長い．

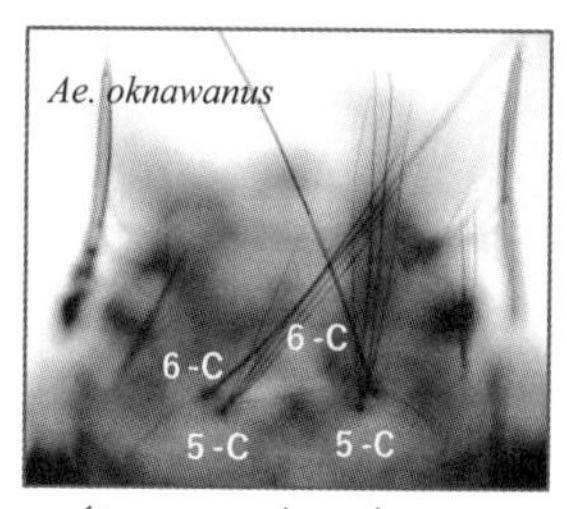

（24ページAへ）

触角毛1(1-A)は分岐する．頭毛5(5-C)と頭毛6(6-C)は頭毛7(7-C)より上でほぼ同じ高さにある．

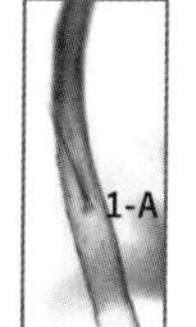

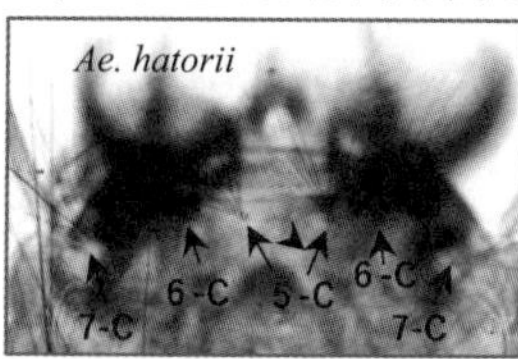

（23ページAへ）

触角毛1(1-A)は通常は単条(ごく稀に2分岐)．頭毛5(5-C)は頭毛6(6-C)より明確に下にある．

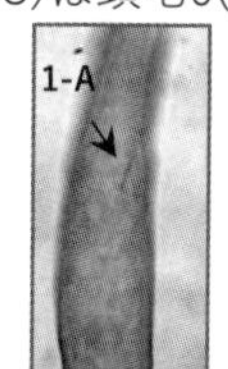

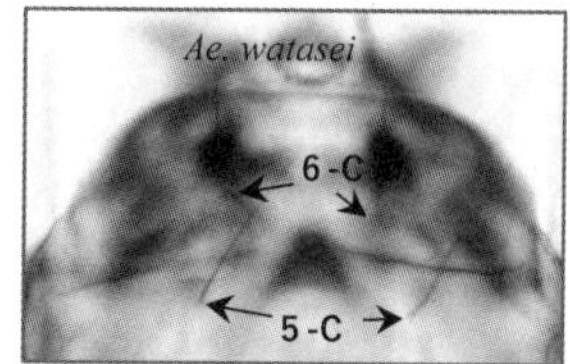

（23ページBへ）

- 22 -

A（22ページより）

B（22ページより）

中胸毛1(1-M)は非常に太く毛根部は隆起する．呼吸管の先端部分は細くなる．呼吸管毛1(1-S)は呼吸管先端より1/2から1/3内にある．鞍板毛1(1-X)は鞍板上にある．

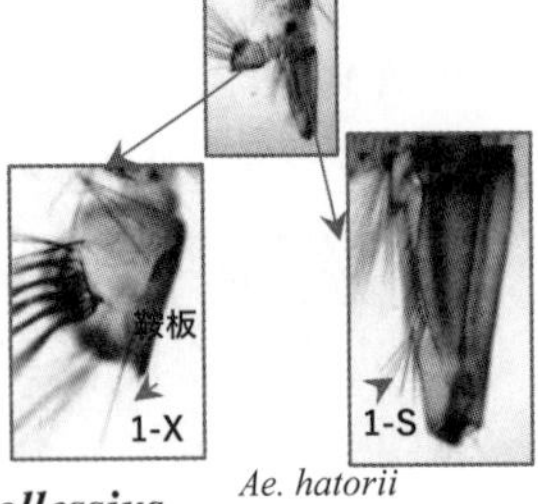

Ae. hatorii

ハトリヤブカ亜属 *Collessius*

ハトリヤブカ *Ae. (Col.) hatorii*

中胸毛1(1-M)は細くて小さい．呼吸管の両側はほぼ平行する．呼吸管毛1(1-S)は呼吸管先端より1/3から1/6内にある．鞍板毛1(1-X)は鞍板上にない．

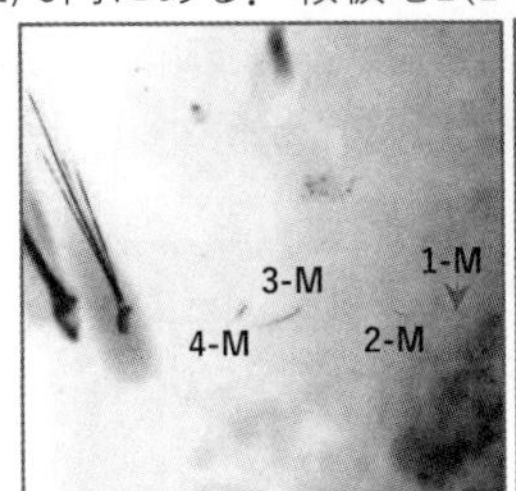

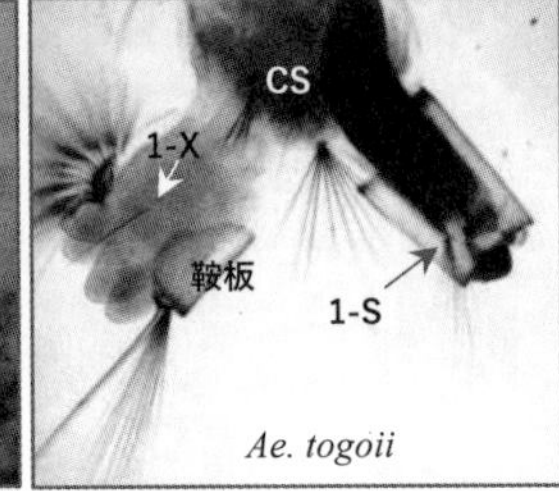

トウゴウヤブカ亜属 *Tanakaius*

第Ⅶ腹節毛3(3-Ⅶ)は第Ⅶ腹節毛4(4-Ⅶ)より長く太い．下唇板の歯は30～36歯．呼吸管には通常明瞭な短横線を前面に認める．側鱗(CS)の先端は太まりヒレ状．

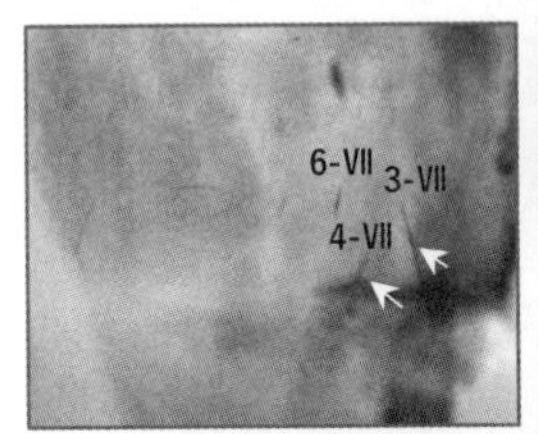

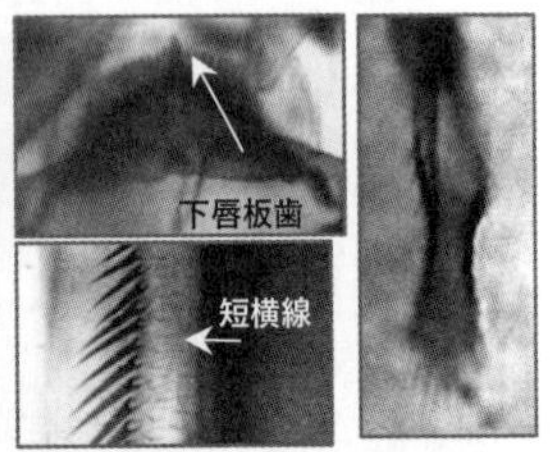

CS

トウゴウヤブカ *Ae. (Tan.) togoi*

第Ⅶ腹節毛3(3-Ⅶ)は第Ⅶ腹節毛4(4-Ⅶ)より短い．下唇板の歯は20～23歯．呼吸管の短横線は明瞭でない．側鱗の先端は太まらず楕円形．

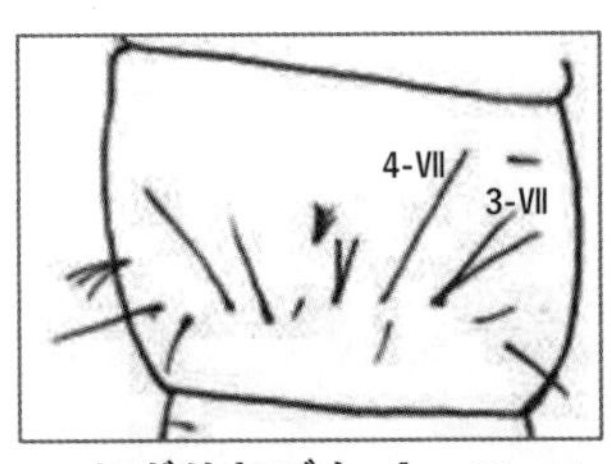

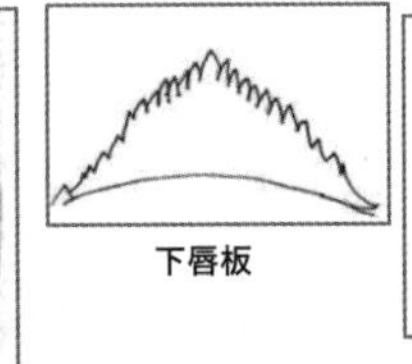

下唇板

CS

セボリヤブカ *Ae. (Tan.) sevoryi*

ワタセヤブカ亜属 *Phagomyia*

頭毛4(4-C)と6(6-C)は頭毛7(7-C)よりやや前方で同じ高さにある．頭毛6(6-C)は2～4分岐．前胸毛1～3(1-P～3-P)は個々に生え，前胸毛1(1-P)は前胸毛3(3-P)より明らかに長い．

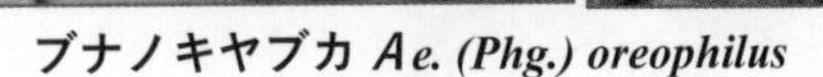

ブナノキヤブカ *Ae. (Phg.) oreophilus*

頭毛4(4-C)は頭毛7(7-C)よりやや前方で頭毛6(6-C)のほぼ真下にある．頭毛6(6-C)は単条．前胸毛1～3(1-P～3-P)は同一肥厚板(硬皮板)上にあり，前胸毛1(1-P)は前胸毛3(3-P)とほぼ同長．

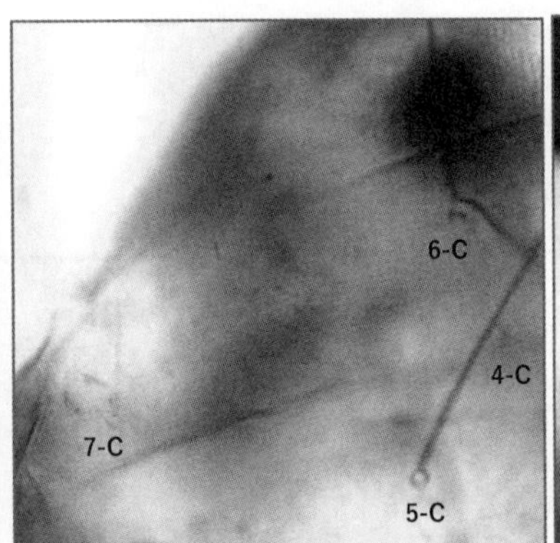

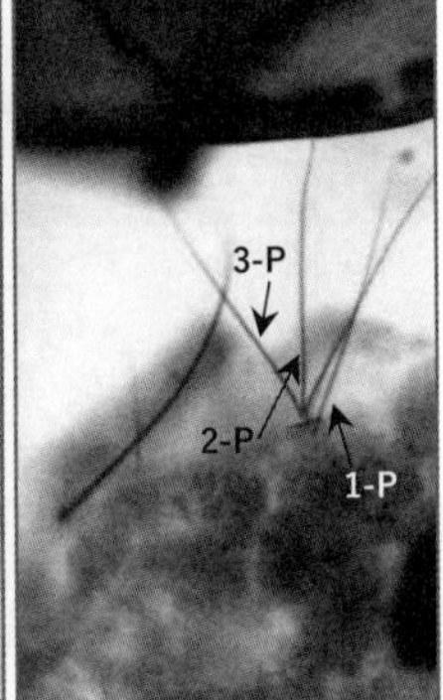

ワタセヤブカ *Ae. (Phg.) watasei*

A（22ページより）

オキナワヤブカ亜属 *Bruceharrisonius* の一部

B（22ページより）

頭毛6(6-C)は頭毛5(5-C)の2倍長以上．側鱗(CS)は45〜59個．第III〜第V腹節毛13(13-III〜V)は第III〜第V腹節毛10(10-III〜V)より長く1〜4分岐．

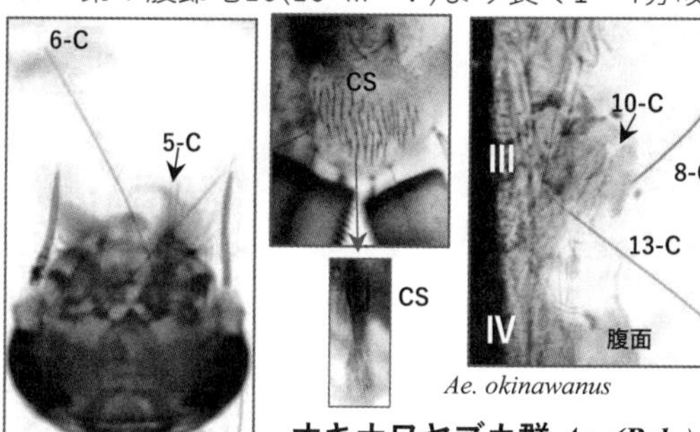

Ae. okinawanus

オキナワヤブカ群 *Ae. (Brh.) okinawanus*

頭毛6(6-C)は頭毛5(5-C)の2倍長未満．側鱗(CS)は28〜38個．第III〜第V腹節毛13(13-III〜V)は第III〜第V腹節毛10(10-III〜V)より短く4〜8分岐．

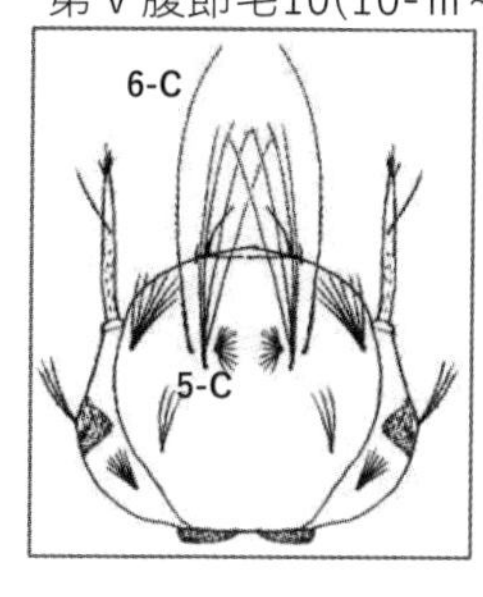

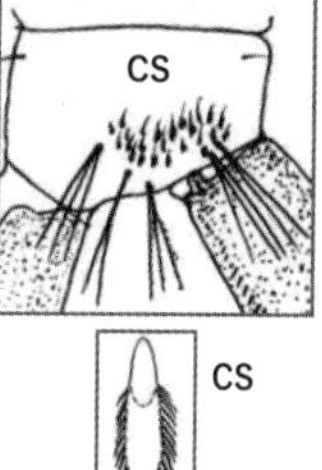

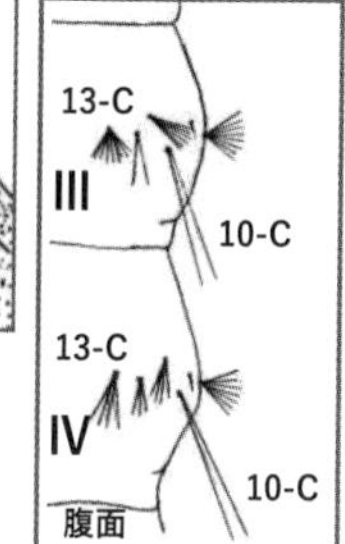

コバヤシヤブカ *Ae. (Brh.) kobayashii*

前胸毛7(7-P)は2分岐．第III腹節毛1(1-III)は通常2(1〜3)分岐．第IV腹節毛1(1-IV)は通常2(1〜2)分岐．

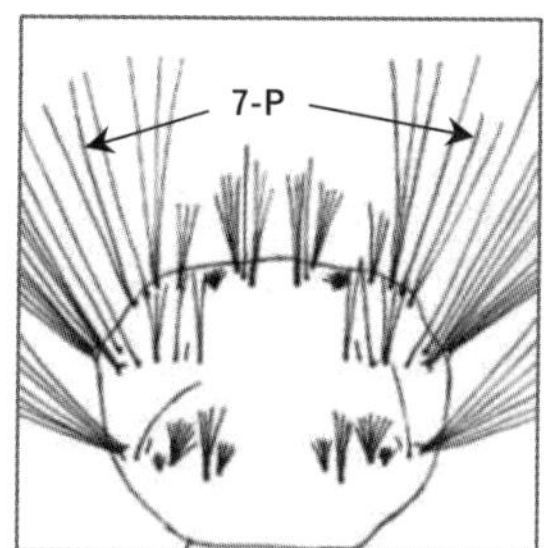

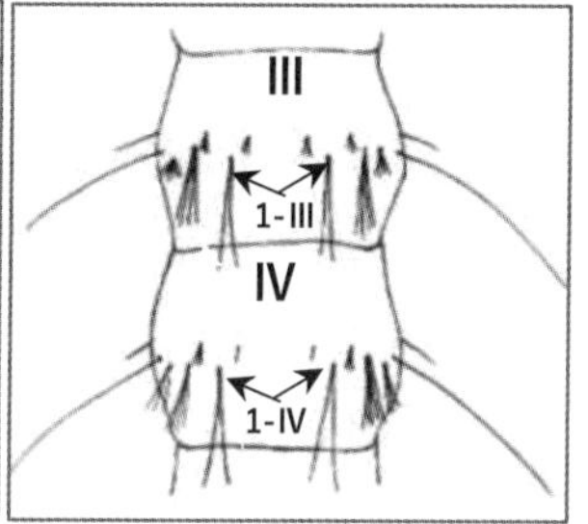

ヤエヤマヤブカ *Ae. (Brh.) okinawanus taiwanus*

前胸毛7(7-P)は3(2〜4)分岐．第III，第IV腹節毛1(1-III,IV)は通常単条(1〜2分岐)．

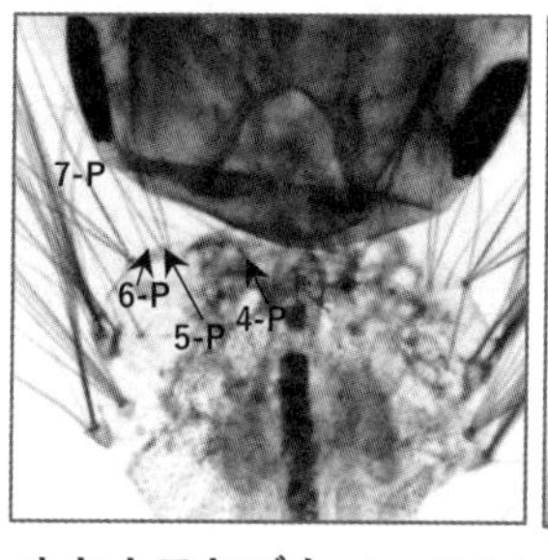

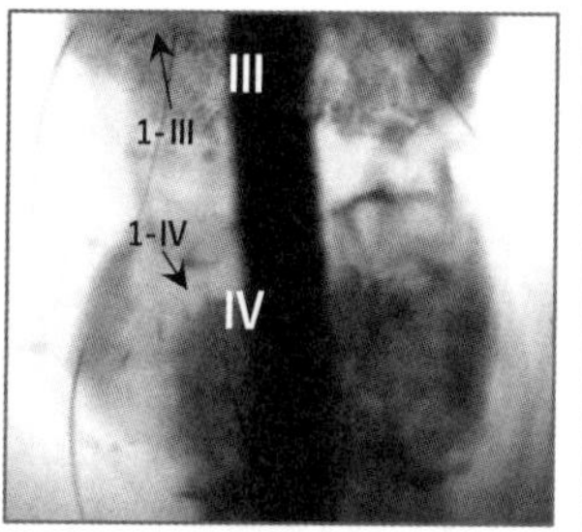

オキナワヤブカ *Ae. (Brh.) okinawanus okinawanus*

側鱗(CS)はほぼ30〜60個で斑をなす．呼吸管棘(PT)はほぼ鱗状で両側に細い側歯を伴う．

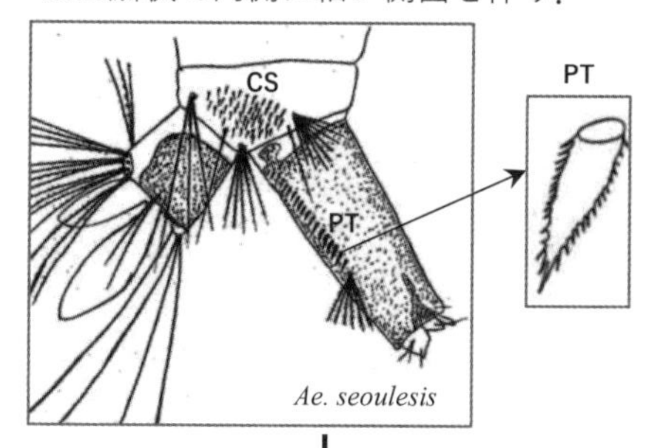

Ae. seoulesis

側鱗(CS)はほぼ6〜17個で列をなす．呼吸管棘(PT)は棘状で腹側に太い側歯を伴う．

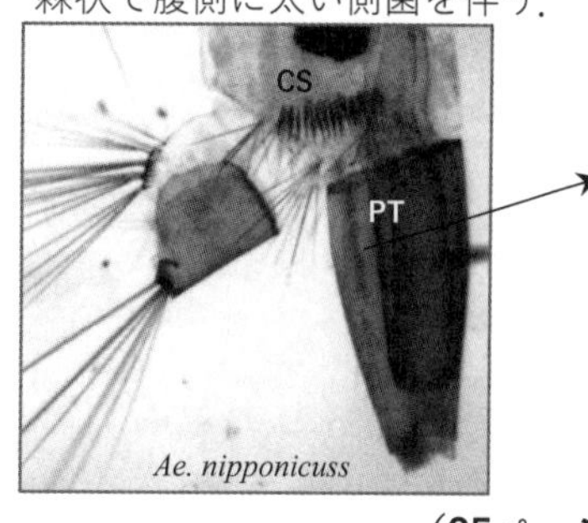

Ae. nipponicuss

（25ページAへ）

ケイジョウヤブカ亜属 *Hopokinsius*

触角毛1(1-A)は3〜5分岐．

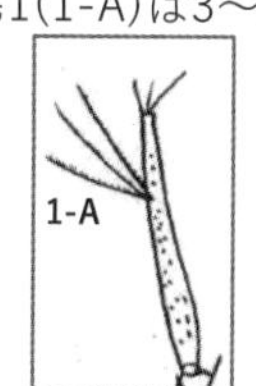

ムネシロヤブカ *Ae. (Hop.) albocinctus*

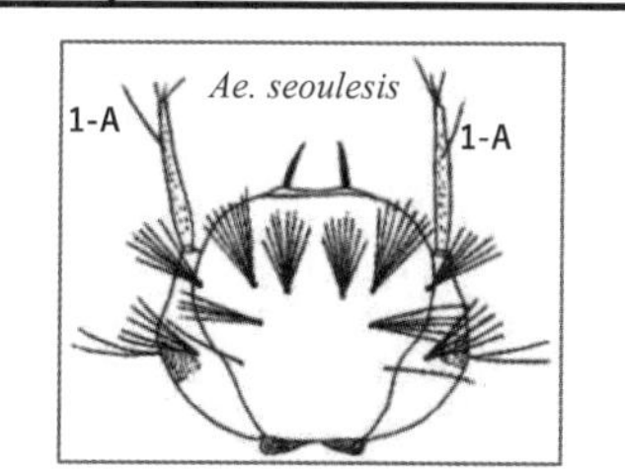

触角毛1(1-A)は単条．

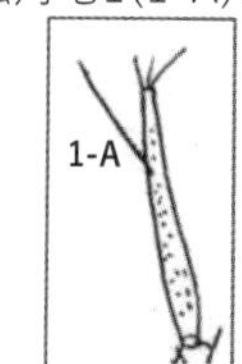

ケイジョウヤブカ *Ae. (Hop.) seoulensis*

A（24ページより）

B（22ページより）

第Ⅰ腹節毛2(2-Ⅰ)は非常に小さく1～2分岐．触角は長く頭長と同じかやや短い．

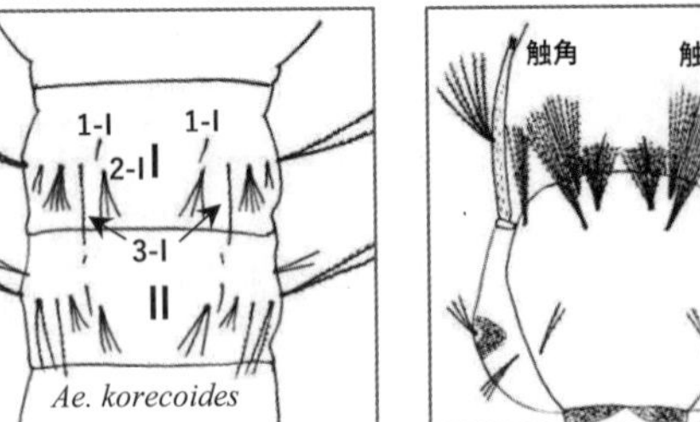

オキナワヤブカ亜属 *Bruceharrionius* の一部

エセチョウセンヤブカ *Ae. (Brh.) koreicoides*

第Ⅰ腹節毛2(2-Ⅰ)は第Ⅰ腹節毛1(1-Ⅰ)とほぼ同長で放射状に3～8分岐．触角は短く頭長の約1/2長．

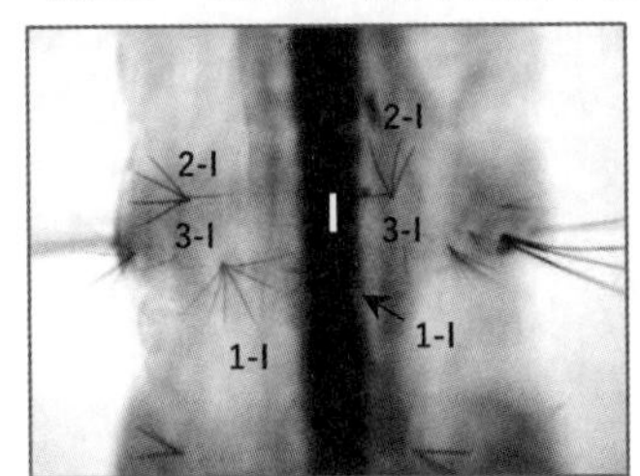

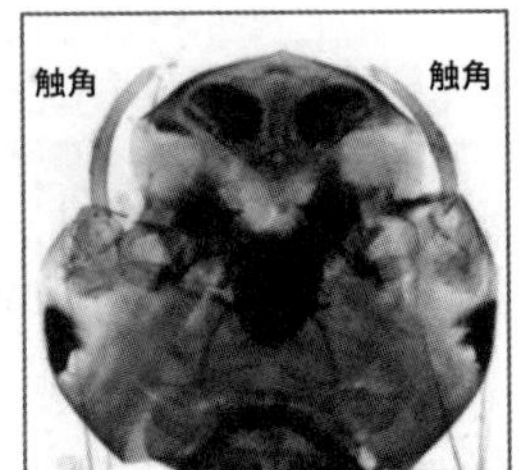

シロカタヤブカ亜属 *Downsiomyia*

尾葉(長い方)は鞍板長の1～1.7倍．

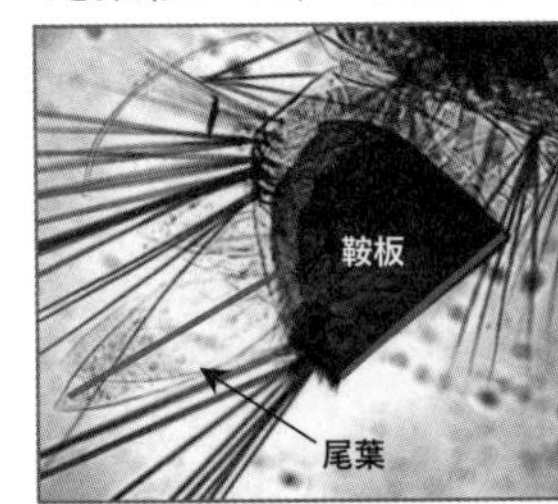

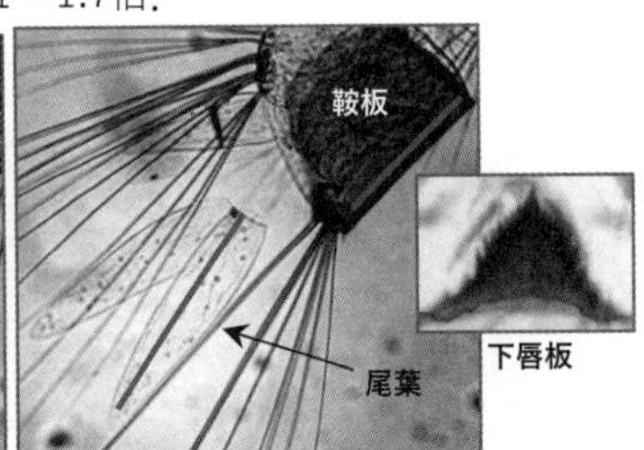

シロカタヤブカ *Ae. (Dow.) nipponicus*

ごくまれに1倍以下のものあり．下唇板歯列の歯はほぼ同型で列の中央に行くに従い縦細にならない．

尾葉(長い方)は鞍板長の0.5～0.7倍．

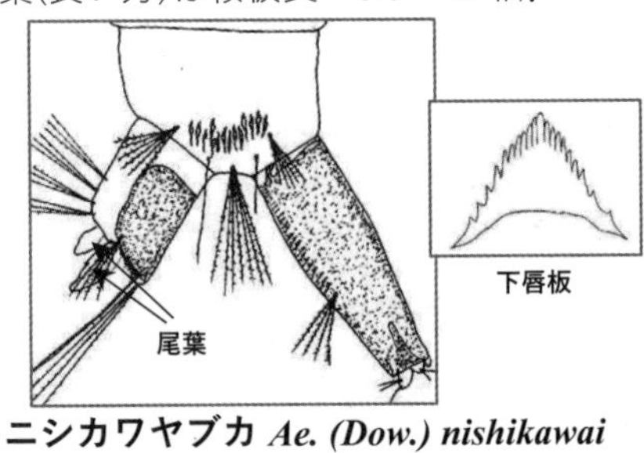

ニシカワヤブカ *Ae. (Dow.) nishikawai*

（トカラ列島中之島，口之島と奄美大島に生息）

頭毛5と6(5-,6-C)は太くなく特に先端部は糸状．

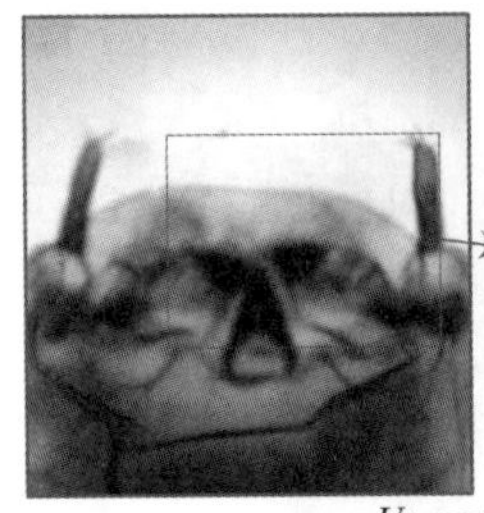

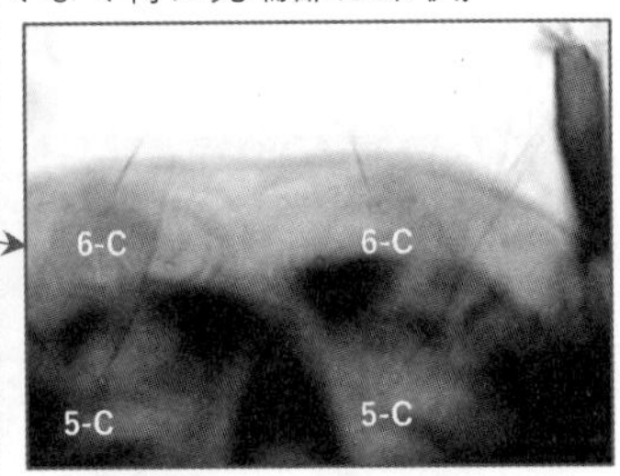

Ur. novobscura

フタクロホシチビカ亜属 *Pseudoficalbia*

頭毛5と6(5-,6-C)は太く黒色芒状．

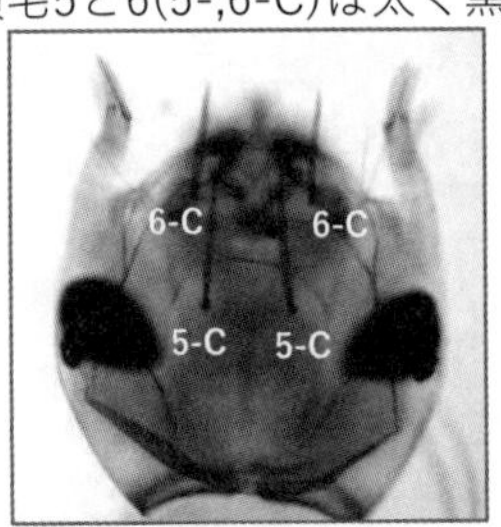

Ur. annandalei

チビカ亜属 *Uranotaenia*

（27ページAへ）

頭毛1(1-C)は痕跡程度．第Ⅳ，第Ⅴ腹節毛5(5-Ⅳ,Ⅴ)は短く弱々しい．第Ⅷ腹節の肥厚板は明瞭．

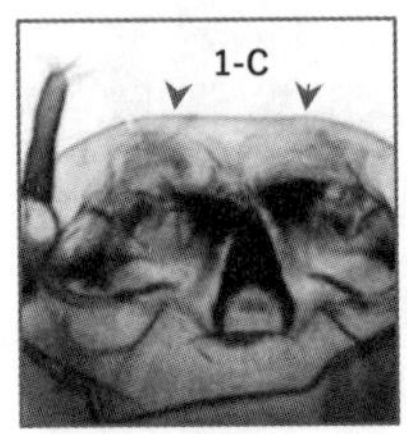

Ur. novobscura

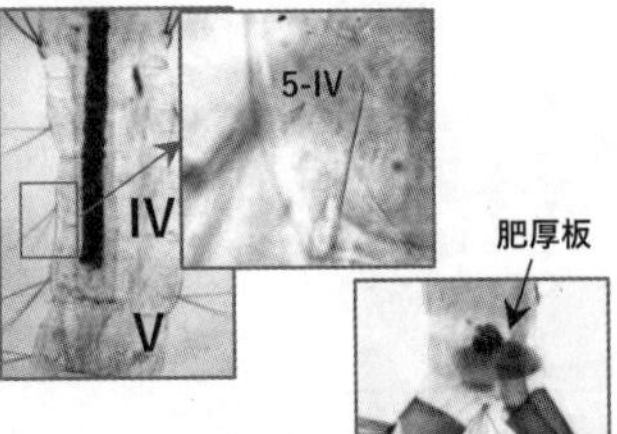

（26ページAへ）

頭毛1(1-C)はへら形．第Ⅳ，第Ⅴ腹節毛5(5-Ⅳ,Ⅴ)は長く強靭．第Ⅷ腹節の肥厚板は不明瞭．

1-C

肥厚板

5-Ⅳ

5-Ⅴ

Ur. yaeyamana

（26ページBへ）

A（25ページより）

B（25ページより）

呼吸管棘は吸管長の1/2より先端側に存在し呼吸管毛(1-S)を越えて生える．頭毛4(4-C)は1~6分岐．第Ⅷ腹節の肥厚板は大きく腹節幅の約7/10．側鱗(CS)は紡錘形．

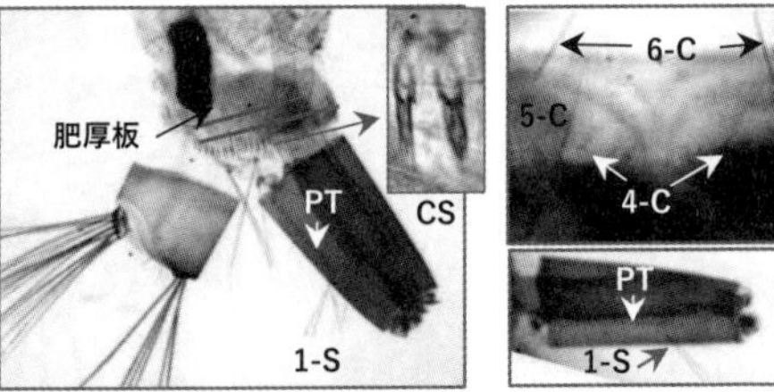

フタクロホシチビカ *Ur. (Pfc.) novobscura*

呼吸管棘(PT)は吸管長の1/2より基部側に存在し呼吸管毛(1-S)を越えて生えない．頭毛4(4-C)は単条．第Ⅷ腹節の肥厚板は普通で腹節幅の約1/2．側鱗(CS)はへら状．

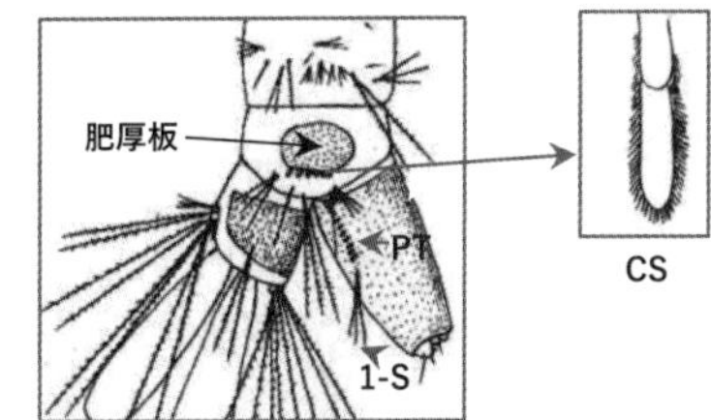

ムネシロチビカ *Ur. (Pfc.) nivipleura*

前胸毛3(3-P)は1～3分岐，4は1～3分岐．呼吸管棘(PT)は18～32．

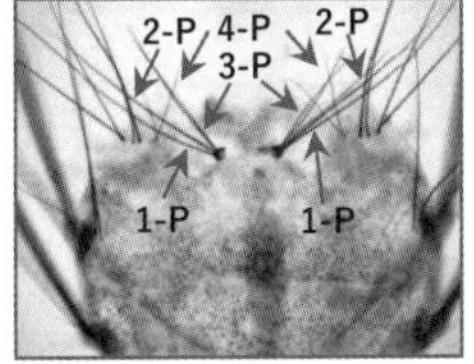

フタクロホシチビカ *Ur. (Pfc.) n. novobscura*

前胸毛3(3-P)は2～5分岐，4は2～6分岐．呼吸管棘(PT)は15～29．

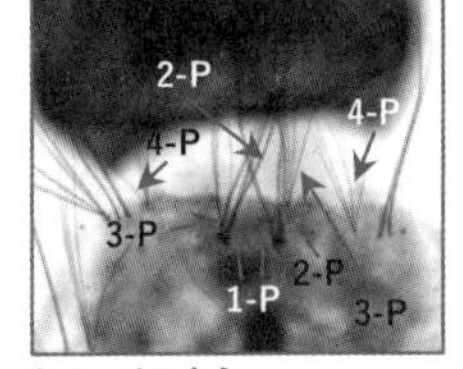

リュウキュウクロホシチビカ *Ur. (Pfc.) n. ryukyuana*

呼吸管棘(PT)は刷毛状ですべて同形．第Ⅲ腹節毛5(5-Ⅲ)は第Ⅳ，第Ⅴ腹節毛5(5-Ⅳ,Ⅴ)と同長．

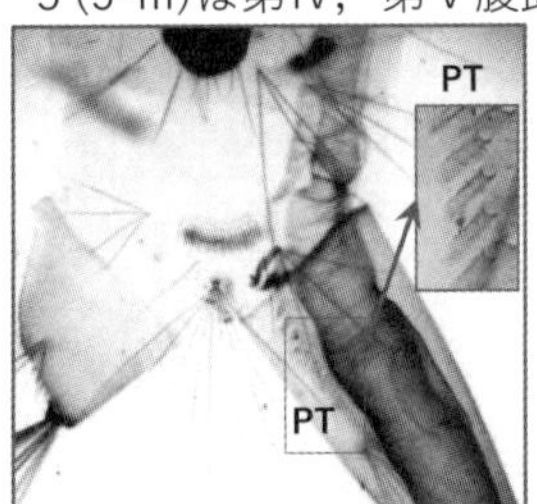

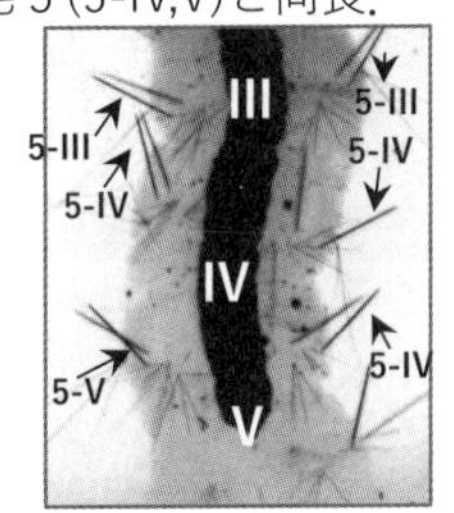

シロオビカニアナチビカ *Ur. (Pfc.) ohamai*

呼吸管棘(PT)は2型を有し先端棘は針状．第Ⅲ腹節毛5(5-Ⅲ)は第Ⅳ，第Ⅴ腹節毛5(5-Ⅳ,Ⅴ)より短い．

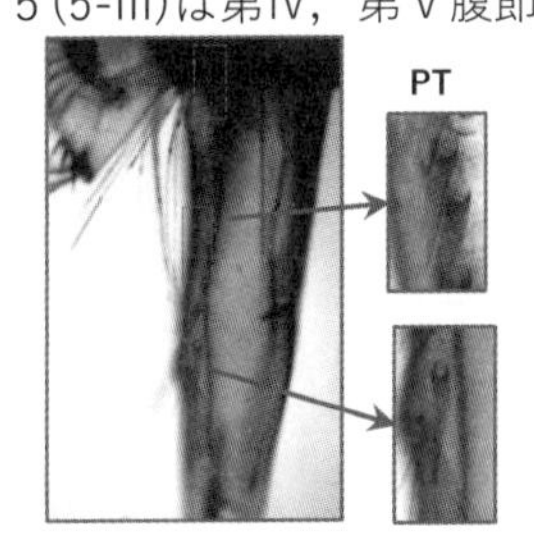

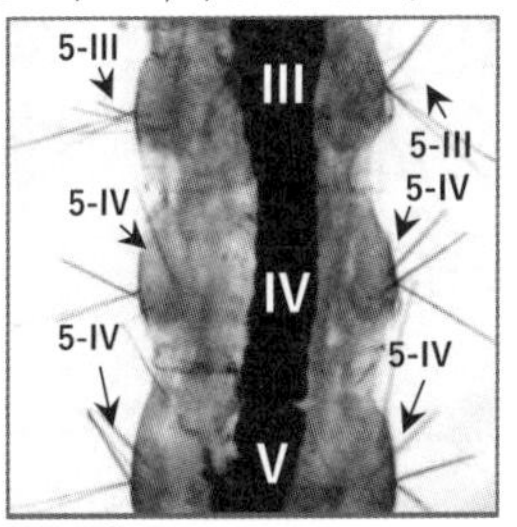

Ur. yaeyamana

下唇板の歯は先端がやや尖り長い．呼吸管毛1(1-S)は呼吸管基部幅とほぼ同長．

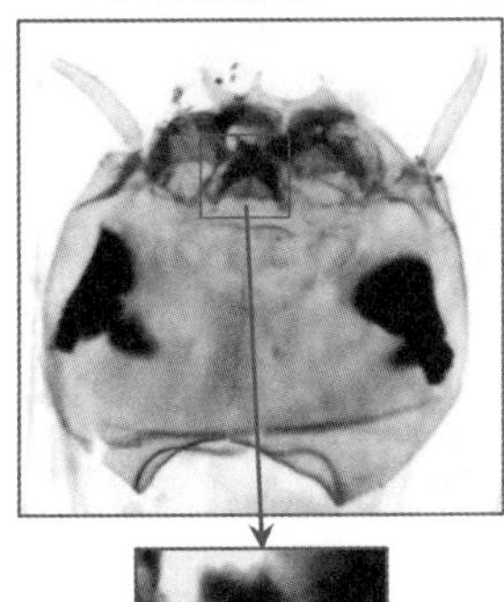

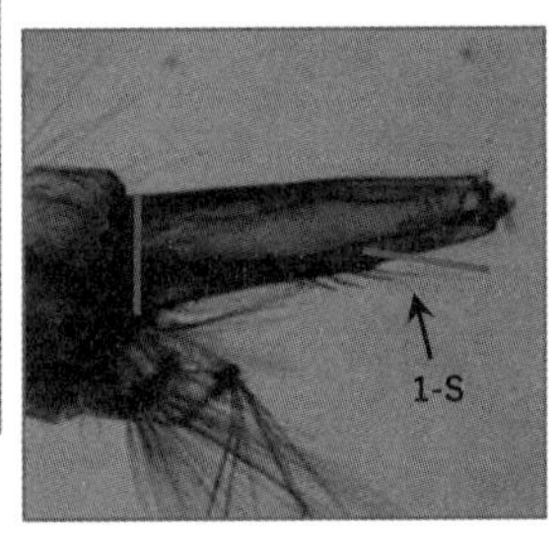

下唇板

ハラグロカニアナチビカ *Ur. (Pfc.) yaeyamana*

下唇板の歯は先端がやや丸く短い．呼吸管毛1(1-S)は呼吸管基部幅より長い．

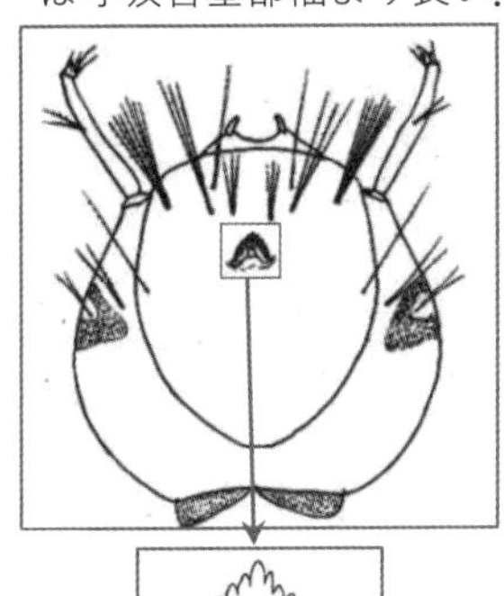

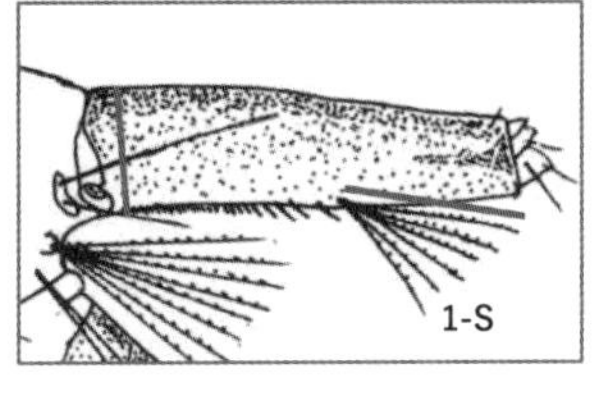

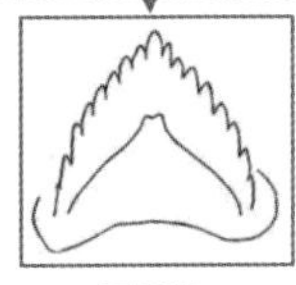

下唇板

カニアナチビカ *Ur. (Pfc.) jacksoni*
（ストウンチビカ）

A（25ページより）

チビカ亜属 *Uranotaenia*

触角の先端には2枚の葉状の触角毛を有する．第Ⅷ腹節の左右の肥厚板は背面で繋がる．側鱗(CS)は4～6個．

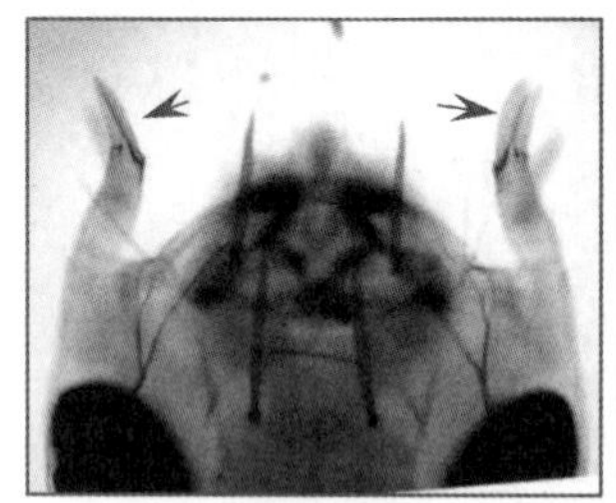

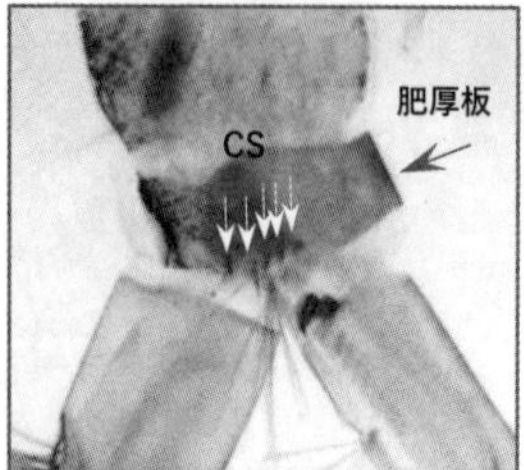

オキナワチビカ *Ur. (Ura.) annandalei*

触角の先端には葉状の触角毛はなく，普通の触角毛を有す．第Ⅷ腹節の左右の肥厚板は背面で繋がらない．側鱗(CS)は6～10個．

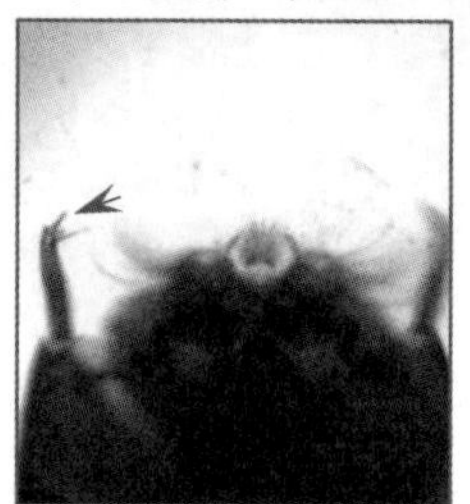

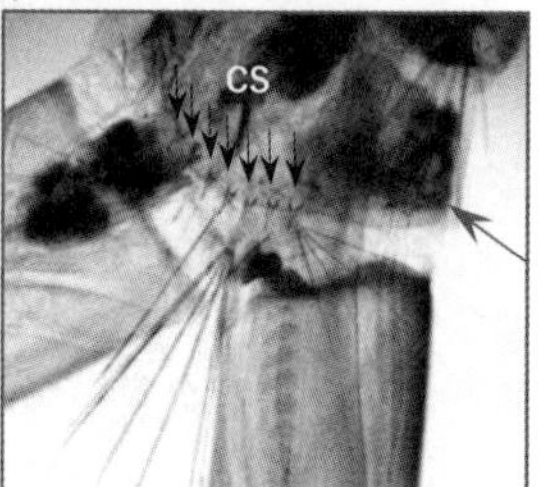

Ur. macfarlanei

呼吸管棘(PT)は通常刷毛状で先端は丸みを帯びず呼吸管毛1(1-S)を超えて存在する．尾葉は丸みを帯び鞍板より遥かに短い．

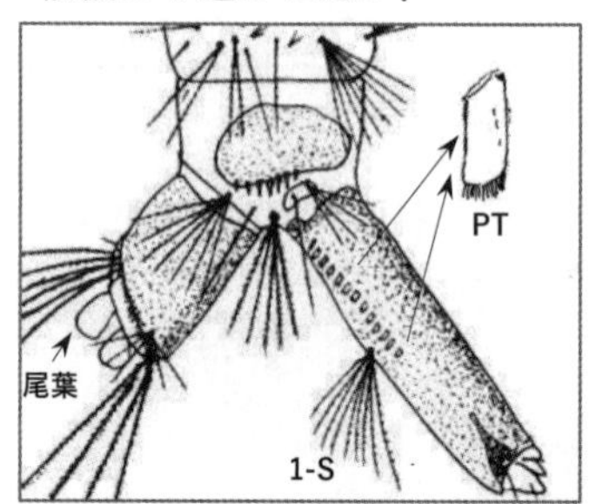

コガタチビカ *Ur. (Ura.) lateralis*

呼吸管棘(PT)は丸みを帯びた刷毛状で吸管毛1(1-S)と少なくとも同位置で超えて存在しない．尾葉は先端が細長く通常鞍板とほぼ同長.

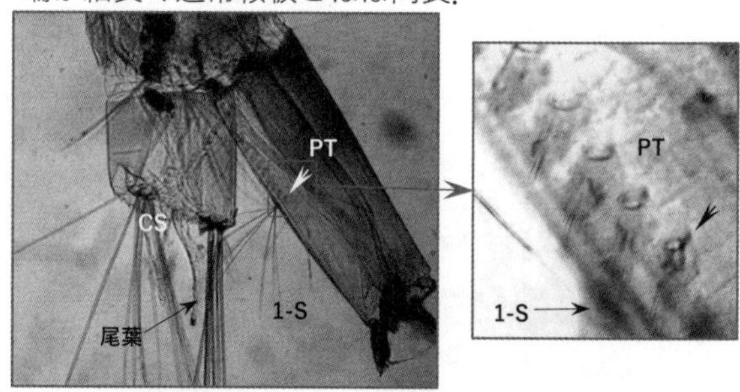

マクファレンチビカ *Ur. (Ura.) macfarlanei*

付表　Appendix

1．日本で生息の記録がある蚊 124 種の種名と学名
Japanese and scientific names of 124 mosquito species recorded from Japan.

2．琉球列島で生息の記録がある蚊とその地理的分布
Geographical distribution of mosquitoes recorded from the Ryukyu Archipelago.

3．琉球列島で生息が記録された蚊の生息水域
Mosquito habitats recorded from the Ryukyu Archipelago.

4．琉球列島産蚊の吸血源動物
Blood-sucking sources of mosquitoes recorded from the Ryukyu Archipelago.

付表1　日本で生息記録がある蚊124種の種名，学名

Appendix 1. Japanese and scientific names of 124 mosquito species recorded from Japan.

	種名 Japanese name　　学　名 Scientific name
1*	モンナシハマダラカ　*Anopheles* (*Anopheles*) *bengalensis* Puri, 1930
2	エンガルハマダラカ　*An.* (*Ano.*) *engarensis* Kanda and Oguma, 1978
3	チョウセンハマダラカ　*An.* (*Ano.*) *koreicus* Yamada and Watanabe, 1918
4*	オオツルハマダラカ　*An.* (*Ano.*) *lesteri* Baisas and Hu, 1936
5*	ヤマトハマダラカ　*An.* (*Ano.*) *lindesayi japonicus* Yamada, 1918
6	オオモリハマダラカ　*An.* (*Ano.*) *omorii* Sakakibara, 1959
7*	オオハマハマダラカ　*An.* (*Ano.*) *saperoi* Bohart and Ingram, 1946
8*	シナハマダラカ　*An.* (*Ano.*) *sinensis* Wiedemann, 1828
9	エセシナハマダラカ　*An.* (*Ano.*) *sineroides* Yamada, 1924
10	ヤツシロハマダラカ　*An.* (*Ano.*) *yatsushiroensis* Miyazaki, 1951
11*	タテンハマダラカ　*An.* (*Cellia*) *tessellatus* Theobald, 1901
12*	コガタハマダラカ（ヤエヤマコガタハマダラカ） *An.* (*Cel.*) *yaeyamaensis* Somboon and Harbach, 2010
13	エゾヤブカ　*Aedes* (*Aedes*) *esoensis* Yamada, 1921
14	ホッコクヤブカ　*Ae.* (*Aed.*) *sasai* Tanaka, Mizusawa and Saugstad, 1979
15	アカエゾヤブカ　*Ae.* (*Aed.*) *yamadai* Sasa, Kano and Hayashi, 1950
16	オオムラヤブカ　*Ae.* (*Aedimorphus*) *alboscutellatus* (Theobald, 1905)
17*	キンイロヤブカ　*Ae.* (*Adm.*) *vexans nipponii* (Theobald, 1907)
18	コバヤシヤブカ　*Ae.* (*Bruceharrisonius*) *kobayashii* Nakata, 1956
19	エセチョウセンヤブカ　*Ae.* (*Brh.*) *koreicoides* Sasa, Kano and Hayashi, 1950
20*	オキナワヤブカ　*Ae.* (*Brh.*) *okinawanus okinawanus* Bohart, 1946
21*	ヤエヤマヤブカ　*Ae.* (*Brh.*) *o. taiwanus* Lien, 1968
22	ハトリヤブカ　*Ae.* (*Collessius*) *hatorii* Yamada, 1921
23	シロカタヤブカ　*Ae.* (*Downsiomyia*) *nipponicus* LaCasse and Yamaguti, 1948
24*	ニシカワヤブカ　*Ae.* (*Dow.*) *nishikawai* Tanaka, Mizusawa and Saugstad, 1979
25	コガタキンイロヤブカ　*Ae.* (*Edwardsaedes*) *bekkui* Mogi, 1977
26*	カニアナヤブカ　*Ae.* (*Geoskusea*) *baisasi* Knight and Hull, 1951
27*	ムネシロヤブカ　*Ae.* (*Hopkinsius*) *albocinctus* (Barraud, 1924)
28	ケイジョウヤブカ　*Ae.* (*Hop.*) *seoulensis* Yamada, 1921
29	ヤマトヤブカ　*Ae.* (*Hulecoeteomyia*) *japonicus japonicus* (Theobald, 1901)
30*	アマミヤブカ　*Ae.* (*Hul.*) *j. amamiensis* Tanaka, Mzusawa and Saugstad, 1979
31*	サキシマヤブカ　*Ae.* (*Hul.*) *j. yaeyamensis* Tanaka, Mzusawa and Saugstad, 1979
32*	ナンヨウヤブカ　*Ae.* (*Neomelaniconion*) *lineatopennis* Ludlowd, 1905
33	アッケシヤブカ　*Ae.* (*Ochlerotatus*) *akkeshiensis* Tanaka, 1998
34	トカチヤブカ　*Ae.* (*Och.*) *communis* (Dc Geer, 1776)
35	ヒサゴヌマヤブカ　*Ae.* (*Och.*) *daiantaeus* Howard, Dyar and Knab, 1913
36	セスジヤブカ　*Ae.* (*Och.*) *dorsalis* (Meigen, 1830)
37	アカンヤブカ　*Ae.* (*Och.*) *exerucians* (Walker, 1856)

付表１（続き）

38	ハクサンヤブカ　*Ae.* (*Och.*) *hakusanensis* Yamaguti and Tamaboko, 1954
39	キタヤブカ　*Ae.* (*Och.*) *hakusanensis hokkaidensis* Tanaka, Mzusawa and Saugstad, 1979
40	ダイセツヤブカ　*Ae.* (*Och.*) *impiger daisetsuzanus* Tanaka, Mzusawa and Saugstad, 1979
41	サッポロヤブカ　*Ae.* (*Och.*) *intrudens* Dyar, 1919
42	チシマヤブカ　*Ae.* (*Och.*) *punctor* (Kirby, 1837)
43	カラフトヤブカ　*Ae.* (*Och.*) *sticticus* (Meigen, 1838)
44*	ハマベヤブカ　*Ae.* (*Och.*) *vigilax* (Skuse, 1889)
45	ブナノキヤブカ　*Ae.* (*Phagomyia*) *oreophilus* Edwards, 1961
46*	ワタセヤブカ　*Ae.* (*Phg.*) *watasei* Yamada, 1921
47*	ネッタイシマカ　*Ae.* (*Stegomyia*) *aegypti* (Linnaeus, 1962)
48*	ヒトスジシマカ　*Ae.* (*Stg.*) *albopictus* (Skuse, 1895)
49*	ダイトウシマカ　*Ae.* (*Stg.*) *daitensis* Miyagi and Toma, 1980
50	ヤマダシマカ　*Ae.* (*Stg.*) *flavopictus flavopictus* Yamada, 1921
51*	ダウンスシマカ（ダウンズシマカ）　*Ae.* (*Stg.*) *f. downsi* Bohart and Ingram, 1946
52*	ミヤラシマカ　*Ae.* (*Stg.*) *flavopictus miyarai* Tanaka, Mzusawa and Saugstad, 1979
53	ミスジシマカ　*Ae.* (*Stg.*) *galloisi* Yamada, 1921
54*	リバースシマカ（リバーズシマカ）　*Ae.* (*Stg.*) *riversi* Bohart and Ingram, 1946
55	タカハシシマカ　*Ae.* (*Stg.*) *wadai* Tanaka, Mzusawa and Saugstad, 1979
56	セボリヤブカ　*Ae.* (*Tanakaius*) *savoryi* Bohart, 1957
57*	トウゴウヤブカ　*Ae.* (*Tan.*) *togoi* (Theobald, 1907)
58*	オオクロヤブカ　*Armigeres subalbatus* (Coquillett, 1898)
59*	アマミムナゲカ　*Heizmannia* (*Heizmannia*) *kana* Tanaka, Mizusawa and Suagstad, 1979
60*	コガタフトオヤブカ　*Verrallina* (*Harbachius*) *nobukonis* (Yamada, 1932)
61*	アカフトオヤブカ　*Ve.* (*Neomacleaya*) *atriisimilis* (Tanaka and Mizusawa, 1973)
62*	クロフトオヤブカ　*Ve.* (*Verrallina*) *iriomotensis* (Tanaka and Mizusawa, 1973)
63	イナトミシオカ　*Culex* (*Barraudius*) *inatomii* Kamimura and Wada, 1974
64*	オビナシイエカ　*Cx.* (*Culex*) *fuscocephala* Theobald, 1907
65*	ジャクソンイエカ　*Cx.* (*Cux.*) *jacksoni* Edwards, 1934
66*	ミナミハマダライエカ　*Cx.* (*Cux.*) *mimeticus* Noé, 1899
67	ハマダライエカ　*Cx.* (*Cux.*) *orientalis* Edwards, 1921
68	アカイエカ　*Cx.* (*Cux.*) *pipiens pallens* Coquillett, 1898
69	チカイエカ　*Cx.* (*Cux.*) *pipiens* form *molestus* Forskái, 1775
70*	シロハシイエカ　*Cx.* (*Cux.*) *pseudovishnui* Colless, 1957
71*	ネッタイイエカ　*Cx.* (*Cux.*) *quinquefasciatus* Say, 1823
72*	ヨツホシイエカ（ヨツボシイエカ）　*Cx.* (*Cux.*) *sitiens* Wiedemann, 1828
73*	コガタアカイエカ　*Cx.* (*Cux.*) *tritaeniorhynchus* Giles, 1901
74*	スジアシイエカ　*Cx.* (*Cux.*) *vagans* Wiedemann, 1828
75*	ニセシロハシイエカ　*Cx.* (*Cux.*) *vishnui* Theobald, 1901
76*	セシロイエカ（セジロイエカ）　*Cx.* (*Cux.*) *whitmorei* (Giles, 1904)
77	キョウトクシヒゲカ　*Cx.* (*Culiciomyia*) *kyotoensis* Yamaguti and LaCasse, 1952

付表1（続き）

78*	クロフクシヒゲカ *Cx.* (*Cul.*) *nigropunctatus* Edwards, 1926
79*	アカクシヒゲカ *Cx.* (*Cul.*) *pallidothorax* Theobald, 1905
80*	リュウキュウクシヒゲカ *Cx.* (*Cul.*) *ryukyensis* Bohart, 1946
81	ヤマトクシヒゲカ *Cx.* (*Cul.*) *sasai* Kano, Nitahara and Awaya, 1954
82*	カギヒゲクロウスカ *Cx.* (*Eumelanomyia*) *brevipalpis* (Giles, 1902)
83	コガタクロウスカ *Cx.* (*Eum.*) *hayashii hayashii* Yamada, 1946
84*	リュウキュウクロウスカ *Cx.* (*Eum.*) *h. ryukyuanus* Tanaka, Mizusawa and Saugstad, 1979
85*	オキナワクロウスカ *Cx.* (*Eum.*) *okinawae* Bohart, 1953
86*	クロツノフサカ *Cx.* (*Lophoceraomyia*) *bicornutus* (Theobald, 1910)
87*	ハラオビツノフサカ *Cx.* (*Lop.*) *cinctellus* Edwards, 1922
88*	フトシマツノフサカ *Cx.* (*Lop.*) *infantulus* Edwards, 1922
89*	アカツノフサカ *Cx.* (*Lop.*) *rubithoracis* (Leicester, 1908)
90*	カニアナツノフサカ *Cx.* (*Lop.*) *tuberis* Bohart. 1946
91	エゾウスカ *Cx.* (*Neoculex*) *rubensis* Sasa and Takahashi. 1948
92*	カラツイエカ *Cx.* (*Oculeomyia*) *bitaeniorhynchus* Giles, 1901
93*	ミツホシイエカ *Cx.* (*Ocu.*) *sinensis* Theobald, 1903
94	オガサワライエカ *Cx.* (*Sirivanakarnius*) *boninensis* Bohart, 1957
95	シノナガカクイカ *Lutzia* (*Insulalutzia*) *shinonagai* (Tanaka, Mizusawa and Suagstad, 1979)
96*	サキジロカクイカ *Lt.* (*Metalutzia*) *fuscana* (Wiedemann, 1820)
97*	トラフカクイカ *Lt.* (*Mlt.*) *vorax* Edwards, 1921
98	ヤマトハボシカ *Culiseta* (*Culicella*) *nipponica* LaCasse and Yamaguti, 1950
99	ミスジハボシカ *Cs.* (*Culiseta*) *kanayamensis* (Yamada, 1932)
100*	オキナワエセコブハシカ *Ficalbia ichiromiyagii* Toma and Higa, 2004
101*	マダラコブハシカ *Mimomyia* (*Etorleptiomyia*) *elegans* (Taylor, 1914)
102*	ルソンコブハシカ *Mi.* (*Eto.*) *luzonensis* (Ludlow, 1905)
103*	ムラサキヌマカ *Coquillettidia* (*Coquillettidia*) *crassipes* (Van der Wulp, 1881)
104*	キンイロヌマカ *Cq.* (*Coq.*) *ochracea* (Theobald, 1903)
105*	アシマダラヌマカ *Mansonia* (*Mansonioides*) *uniformis* (Theobald, 1901)
106*	ハマダラナガスネカ *Orthopodomyia anopheloides* (Giles, 1903)
107*	オキナワカギカ *Malaya genurostris* Leicester, 1908
108*	ヤンバルギンモンカ *Topomyia* (*Suaymyia*) *yanbarensis* Miyagi, 1976
109	キンパラナガハシカ *Tripteroides* (*Tripteroides*) *bambusa* (Yamada, 1917)
110*	ヤエヤマナガハシカ *Tp.* (*Trp.*) *yaeyamensis* Tanaka, Mizusawa and Saugstad, 1979
111*	ヤエヤマオオカ *Toxorhynchites* (*Toxorhynchites*) *manicatus yaeyamae* Bohart, 1956
112*	ヤマダオオカ *Tx.* (*Tox.*) *m. yamadai* (Ôuchi, 1939)
113*	オキナワオオカ *Tx.* (*Tox.*) *okinawensis* Toma, Miyagi and Tanaka, 1990
114	トワダオオカ *Tx.* (*Tox.*) *towadensis* (Matsumura, 1916)
115*	カニアナチビカ *Uranotaenia* (*Pseudoficalbia*) *jacksoni* Edwards, 1935
116*	ムネシロチビカ *Ur.* (*Pfc.*) *nivipleura* Leicester, 1908
117	フタクロホシチビカ *Ur.* (*Pfc.*) *novobscura novobscura* Barraud, 1934

付表１（続き）

118*	リュキュウクロホシチビカ *Ur.* (*Pfc.*) *novobscura ryukyuana* Tanaka, Mizusawa and Saugstad, 1979
119*	シロオビカニアナチビカ *Ur.* (*Pfc.*) *ohamai* Tanaka, Mizusawa and Saugstad, 1975
120*	イリオモテチビカ *Ur.* (*Pfc.*) *tanakai* Miyagi and Toma, 2013
121*	ハラグロカニアナチビカ *Ur.* (*Pfc.*) *yaeyamana* Tanaka, Mizusawa and Saugstad, 1975
122*	オキナワチビカ *Ur.* (*Uranotaenia*) *annandalei* Barraud, 1926
123*	コガタチビカ *Ur.* (*Ura.*) *lateralis* Ludlow, 1905
124*	マクファレンチビカ *Ur.* (*Ura.*) *macfarlanei* Edwards, 1914

・付表中の太文字でアスタリック (*) がついた数字は琉球列島で記録がある蚊の種を示す（77 種）

・The bold numbers with asterisks (*) in the appendix indicate mosquito species recorded from the Ryukyu Islands (77 species).

付表2　琉球列島で生息の記録がある蚊とその地理的分布

種名	琉球列島																			東洋区（含台湾）	旧北区日本
	中之島	宝島	口之島	奄美大島	徳之島	沖縄島	伊平屋島	伊是名島	水納島	久米島	宮古島	石垣島	西表島	黒島	波照間島	小浜島	与那国島	北大東島	南大東島		
モンナシハマダラカ				○	○															○	
オオツルハマダラカ				○		○	○					○	○				○			○	○
ヤマトハマダラカ	○	○																			○
オオハマハマダラカ						○						○	○								
シナハマダラカ	○	○	○	○	○	○	○	○		○	○	○	○	○	○	○	○	○	○	○	○
タテンハマダラカ						○						○	○				○			○	
コガタハマダラカ											○	○	○			○	?				
キンイロヤブカ	○			○	○	○	○		○		○	○	○		○		○		○		○
オキナワヤブカ				○	○	○				○											○
ヤエヤマヤブカ												○	○							○	
ニシカワヤブカ	○			○	○																
カニアナヤブカ				○		○			○		○	○	○				○			○	
ムネシロヤブカ													○							○	
アマミヤブカ				○	○	?															
サキシマヤブカ												○	○								
ナンヨウヤブカ											○	○	○				○			○	
ハマベヤブカ														○						○	
ワタセヤブカ				○	○	○						○	○								○
ネッタイシマカ						○				○	○	○	○							○	
ヒトスジシマカ	○	○	○	○	○	○	○	○	○	○	○	○	○	○	○	○	○	○	○	○	○
ダイトウシマカ																		○	○		
ダウンスシマカ	○			○	○	○				○											
ミヤラシマカ												○	○								
リバースシマカ	○	○		○	○	○	○			○	◎	○	○	○	○		○				○
トウゴウヤブカ	○	○	○	○	○	○	○		○	○	○	○	○	○	○		○	○	○	○	○
オオクロヤブカ	○	○	○	○	○	○	○	○	○	○	○	○	○	○	○	○	○	○	○	○	○
アマミムナゲカ				○	○																
コガタフトオヤブカ						○						○	○								○
アカフトオヤブカ													○								
クロフトオヤブカ													○								
オビナシイエカ												○	○			○	○			○	
ジャクソンイエカ						○														○	○
ミナミハマダライエカ	○			○	○	○	○	○		○	○		○							○	○
シロハシイエカ				○	○	○	○	○		○	○	○	○				○			○	○
ネッタイイエカ	○	○		○	○	○	○	○	○	○	○	○	○	○	○	○	○	○	○	○	○
ヨツホシイエカ						○					○	○	○							○	
コガタアカイエカ	○	○	○	○	○	○	○	○	○	○	○	○	○		○	○	○	○	○	○	○
スジアシイエカ				○		○	○										○			○	○
ニセシロハシイエカ						○						○	○							○	

付表 2 （続き）

種名	琉球列島																			東洋区（含台湾）	旧北区日本
	中之島	宝島	口之島	奄美大島	徳之島	沖縄島	伊平屋島	伊是名島	水納島	久米島	宮古島	石垣島	西表島	黒島	波照間島	小浜島	与那国島	北大東島	南大東島		
セシロイエカ				○		○						○	○							○	○
クロフクシヒゲカ												○	○				○			○	
アカクシヒゲカ				○		○			○				○				○			○	○
リュウキュウクシヒゲカ	○			○	○	○	○		○			○	○				○				
カギヒゲクロウスカ						○						○	○							○	
リュウキュウクロウスカ				○	○	○					○	○	○	○							
オキナワクロウスカ				○		○						○	○							○	
クロツノフサカ												○	○				○			○	
ハラオビツノフサカ												○	○				○			○	
フトシマツノフサカ				○	○	○						○	○				○			○	○
アカツノフサカ				○		○	○												○	○	○
カニアナツノフサカ					○	○				○		○	○							○	
カラツイエカ		○		○	○	○	○			○	○	○	○		○	○	○			○	○
ミツホシイエカ				○		○					?	○	○							○	○
サキジロカクイカ						○			○		○	○	○					○	○	○	
トラフカクイカ	○			○	○	○	○	○	○	○	○	○	○	○	○		○		○	○	○
オキナワエセコブハシカ													○								
マダラコブハシカ						○	○					○	○				○			○	
ルソンコブハシカ				○		○	○			○	○	○	○			○	○			○	
ムラサキヌマカ				○	○	○	○				○	○	○				○		○	○	
キンイロヌマカ	○			○		○						○	○							○	○
アシマダラヌマカ				○		○	○	○			○	○	○			○	○		○	○	○
ハマダラナガスネカ	○			○	○	○				○		○	○							○	○
オキナワカギカ	○			○	○	○					?	○	○	○			○		○	○	
ヤンバルギンモンカ	○	○		○	○	○				○		○	○							○	○
ヤエヤマナガハシカ	○											○	○				○				
ヤエヤマオオカ												○	○								
ヤマダオオカ				○	○																
オキナワオオカ						○															
カニアナチビカ				○	○	○				○			?							○	
ムネシロチビカ				○	○	○						○	○							○	
リュウキュウクロホシチビカ				○	○	○	○			○		○	○								
シロオビカニアナチビカ												○	○				○				
イリオモテチビカ												○	○								
ハラグロカニアナチビカ													○								
オキナワチビカ						○						○	○				○			○	
コガタチビカ													○							○	
マクファレンチビカ					○	○						○	○							○	

○は生息の記録有（宮城・當間，2017）；◎は新記録（前川，私信），？は生息が不明

Appendix 2. Geographical distribution of mosquitoes recorded from the Ryukyu Archipelago.

Species	Ryukyu Archipelago																			Oriental R. (inclu. Taiwan)	Palaearctic Japan
	Nakanoshima	Takarajima	Kuchinoshima	Amamioshima	Tokunoshima	Okinawajima	Iheyajima	Izenajima	Minnajima	Kumejima	Miyakojima	Ishigakijima	Iriomotejima	Kuroshima	Haterumajima	Kohamajima	Yonagunijima	kitadaitojima	Minamidaitojima		
An. bengalensis				○	○															○	
An. lesteri				○		○	○					○	○				○			○	○
An. l. japonicus	○	○																			○
An. saperoi						○						○	○								
An. sinensis	○	○	○	○	○	○	○	○		○	○	○	○	○	○	○	○	○	○	○	○
An. tessellatus						○						○	○				○			○	
An. yaeyamaensis											○	○	○			○	?				
Ae. v. nipponii	○			○	○	○	○		○		○	○	○		○		○		○		○
Ae. o. okinawanus				○	○	○				○											○
Ae. o. taiwanus												○	○							○	
Ae. nishikawai	○			○	○																
Ae. baisasi				○		○			○		○	○	○				○			○	
Ae. albocinctus													○							○	
Ae. j. amamiensis				○	○	?															
Ae. j. yaeyamensis												○	○								
Ae. lineatopennis											○	○	○				○			○	
Ae. vigilax														○						○	
Ae. watasei				○	○	○						○	○								○
Ae. aegypti						○				○	○	○	○							○	
Ae. albopictus	○	○	○	○	○	○	○	○	○	○	○	○	○	○	○	○	○	○	○	○	○
Ae. daitensis																		○	○		
Ae. f. downsi	○			○	○	○				○											
Ae. f. miyarai												○	○								
Ae. riversi	○	○		○	○	○	○			○	◎	○	○	○	○		○				○
Ae. togoi	○	○	○	○	○	○	○		○	○	○	○	○	○	○		○	○	○	○	○
Ar. subalbatus	○	○	○	○	○	○	○	○	○	○	○	○	○	○	○	○	○	○	○	○	○
Hz. kana				○	○																
Ve. nobukonis						○						○	○								○
Ve. atriisimilis													○								
Ve. iriomotensis													○								
Cx. fuscocephala												○	○			○	○			○	
Cx. jacksoni						○														○	○
Cx. mimeticus	○			○	○	○	○	○		○	○		○							○	○
Cx. pseudovishnui				○	○	○	○	○		○	○	○	○				○			○	○
Cx. quinquefasciatus	○	○		○	○	○	○	○	○	○	○	○	○	○	○	○	○	○	○	○	○
Cx. sitiens						○					○	○	○							○	
Cx. tritaeniorhynchus	○	○	○	○	○	○	○	○	○	○	○	○	○		○	○	○	○	○	○	○
Cx. vagans				○		○	○										○			○	○
Cx. vishnui						○						○	○							○	

Appendix 2 (continued).

Species	Ryukyu Archipelago																			Oriental R. (inclu. Taiwan)	Palaearctic Japan
	Nakanoshima	Takarajima	Kuchinoshima	Amamioshima	Tokunoshima	Okinawajima	Iheyajima	Izenajima	Minnajima	Kumejima	Miyakojima	Ishigakijima	Iriomotejima	Kuroshima	Haterumajima	Kohamajima	Yonagunijima	kitadaitojima	Minamidaitojima		
Cx. whitmorei				○		○						○	○							○	○
Cx. nigropunctatus												○	○				○			○	
Cx. pallidothorax				○		○			○				○				○			○	○
Cx. ryukyensis	○			○	○	○	○		○			○	○				○				
Cx. brevipalpis						○						○	○							○	
Cx. h. ryukyuanus				○	○	○					○	○	○	○							
Cx. okinawae				○		○						○	○							○	
Cx. bicornutus												○	○				○			○	
Cx. cinctellus												○	○				○			○	
Cx. infantulus				○	○	○						○	○				○			○	○
Cx. rubithoracis				○		○	○												○	○	○
Cx. tuberis					○	○				○		○	○							○	
Cx. bitaeniorhynchus		○		○	○	○	○			○	○	○	○		○	○	○			○	○
Cx. sinensis				○		○					?	○	○							○	○
Lt. fuscana						○			○		○	○	○					○	○	○	
Lt. vorax	○			○	○	○	○	○	○	○	○	○	○	○	○		○		○	○	○
Fi. ichiromiyagii													○								
Mi. elegans						○	○					○	○				○			○	
Mi. luzonensis				○		○	○			○	○	○	○			○	○			○	
Cq. crassipes				○	○	○	○				○	○	○				○		○	○	
Cq. ochracea	○			○		○						○	○							○	○
Ma. uniformis				○		○	○	○			○	○	○			○	○		○	○	○
Or. anopheloides	○			○	○	○				○		○	○							○	○
Ml. genurostris	○			○	○	○					?	○	○	○			○		○	○	
To. yanbarensis	○	○		○	○	○				○		○	○							○	○
Tp. yaeyamensis	○											○	○				○				
Tx. m. yaeyamae												○	○								
Tx. m. yamadai				○	○																
Tx. okinawensis						○															
Ur. jacksoni				○	○	○				○			?							○	
Ur. nivipleura				○	○	○						○	○							○	
Ur. n. ryukyuana				○	○	○	○			○		○	○								
Ur. ohamai												○	○				○				
Ur. tanakai												○	○								
Ur. yaeyamana													○								
Ur. annandalei						○						○	○				○			○	
Ur. lateralis													○							○	
Ur. macfarlanei					○	○						○	○							○	

○ Collection (Miyagi and Toma, 2017); ◎, new record (Maekawa, personal communication);
?, not clear.

付表3　琉球列島で生息が記録された蚊の生息水域

種　名	生息水域																
	山脚および森林地帯											平野部					
	葉腋	竹穴	竹切株	樹洞	人工容器	湿地	地表水	岩礁	渓流	カニ穴（淡水）	カニ穴（塩水）	人工容器	地表水	岩礁	水田	休耕田	タイモ田
モンナシハマダラカ									++								
オオツルハマダラカ							+								+	+	
ヤマトハマダラカ								+	++								
オオハマハマダラカ									++								
シナハマダラカ					+		+		+				+		+++	+++	
タテンハマダラカ															?		
コガタハマダラカ									++								
キンイロヤブカ													++		++	++	
オキナワヤブカ			++	+++	++												
ヤエヤマヤブカ			++	+++	++												
ニシカワヤブカ				++													
カニアナヤブカ											++						
ムネシロヤブカ				?													
アマミヤブカ				++	+			++									
サキシマヤブカ				+++	++			+++									
ナンヨウヤブカ																+	
ハマベヤブカ													+				
ワタセヤブカ			+	++	++												
ネッタイシマカ					+												
ヒトスジシマカ			+	+	++							+++					
ダイトウシマカ				++	++												
ダウンスシマカ			++	+++	+++												
ミヤラシマカ	+		++	+++	+++												
リバースシマカ	+		++	+++	++												
トウゴウヤブカ												+		+++			
オオクロヤブカ	+		++	++	++		+					++	+				
アマミムナゲカ				?													
コガタフトオヤブカ							+										
アカフトオヤブカ							+						+				
クロフトオヤブカ							+										
オビナシイエカ												+	++	+	++	++	
ジャクソンイエカ													+				
ミナミハマダライエカ									+						+	+	
シロハシイエカ													++		++	++	+
ネッタイイエカ					+		+					++	+++			++	
ヨツホシイエカ													++	+		++	
コガタアカイエカ					+		+					+	++		+++	+++	++
スジアシイエカ																+	
ニセシロハシイエカ															+	+	+

付表 3（続き）

種　名	生息水域																
	山脚および森林地帯											平野部					
	葉腋	竹穴	竹切株	樹洞	人工容器	湿地	地表水	岩礁	渓流	カニ穴（淡水）	カニ穴（塩水）	人工容器	地表水	岩礁	水田	休耕田	タイモ田
セシロイエカ													+		+	+	
クロフクシヒゲカ												+++	++		+	++	
アカクシヒゲカ				+	++			+									
リュウキュウクシヒゲカ			+	+++	+++		+++	+++		+			++				
カギヒゲクロウスカ					+												
リュウキュウクロウスカ									++								
オキナワクロウスカ					++		++	++	+								
クロツノフサカ				+++	+++		+	+		+							
ハラオビツノフサカ					+											+	
フトシマツノフサカ					++		++	++	+	+			++			++	
アカツノフサカ															+		
カニアナツノフサカ					+					+	+						
カラツイエカ							+	+				+	++	+	+	++	
ミツホシイエカ																+	
サキジロカクイカ				+	+							++	+	+			
トラフカクイカ			+	+	+		+	+				++	+	+		+	
オキナワエセコブハシカ						+										+	
マダラコブハシカ																++	
ルソンコブハシカ															+	++	
ムラサキヌマカ																+	
キンイロヌマカ																+	
アシマダラヌマカ																+	
ハマダラナガスネカ			++	+++	+++												
オキナワカギカ	++																
ヤンバルギンモンカ		++															
ヤエヤマナガハシカ			++	+++	+++												
ヤエヤマオオカ			++	++	++												
ヤマダオオカ			++	++	++												
オキナワオオカ				+	+												
カニアナチビカ					+					++							
ムネシロチビカ			+														
リュウキュウクロホシチビカ			++	+++	+++			+									
シロオビカニアナチビカ							+		+	++	++						
イリオモテチビカ						+											
ハラグロカニアナチビカ										+							
オキナワチビカ					+		++	+	+								
コガタチビカ													++			+	
マクファレンチビカ							++	+	+	+							

宮城・當間（2017）．+，生息が希；++，普通に生息；+++，多数生息；?，不明

Appendix 3. Mosquito habitats recorded from the Ryukyu Archipelago.

Species	Breeding habitats																
	Mountain hood and forest areas											Open areas					
	Leaf axils	Bamboo holes	Cut bamboos	Tree holes	Artificial containers	Swamps	Ground pools	Rock pools	Streams	Crab holes (fresh water)	Crab holes (brackish water)	Artificial containers	Ground pools	Rock pools	Paddy fields	Unirrigated field	Taro fields
An. bengalensis									++								
An. lesteri							+								+	+	
An. l. japonicus								+	++								
An. saperoi									++								
An. sinensis					+		+		+				+		+++	+++	
An. tessellatus															?		
An. yaeyamaensis									++								
Ae. v. nipponii													++		++	++	
Ae. o. okinawanus			++	+++	++												
Ae. o. taiwanus			++	+++	++												
Ae. nishikawai				++													
Ae. baisasi											++						
Ae. albocinctus				?													
Ae. j. amamiensis				++	+			++									
Ae. j. yaeyamensis				+++	++			+++									
Ae. lineatopennis																+	
Ae. vigilax													+				
Ae. watasei			+	++	++												
Ae. aegypti					+												
Ae. albopictus			+	+	++							+++					
Ae. daitensis				++	++												
Ae. f. downsi			++	+++	+++												
Ae. f. miyarai	+		++	+++	+++												
Ae. riversi	+		++	+++	++												
Ae. togoi												+		+++			
Ar. subalbatus	+		++	++	++		+					++	+				
Hz. kana				?													
Ve. nobukonis							+										
Ve. atriisimilis							+						+				
Ve. iriomotensis							+										
Cx. fuscocephala												+	++	+	++	++	
Cx. jacksoni													+				
Cx. mimeticus									+						+	+	
Cx. pseudovishnui													++		++	++	+
Cx. quinquefasciatus					+		+					++	+++			++	
Cx. sitiens													++	+		++	
Cx. tritaeniorhynchus					+		+					+	++		+++	+++	++
Cx. vagans																+	
Cx. vishnui															+	+	+

Appendix 3. (Continued)

Species	Breeding habitats																
	Mountain hood and forest areas											Open areas					
	Leaf axils	Bamboo holes	Cut bamboos	Tree holes	Artificial containers	Swanps	Ground pools	Rock pools	Streams	Crab holes (fresh water)	Crab holes (brackish water)	Artificial containers	Ground pools	Rock pools	Paddy fields	Unirrigated field	Taro fields
Cx. whitmorei													+		+	+	
Cx. nigropunctatus												+++	++		+	++	
Cx. pallidothorax				+	++			+									
Cx. ryukyensis			+	+++	+++		+++	+++		+			++				
Cx. brevipalpis					+												
Cx. h. ryukyuanus									++								
Cx. okinawae					++		++	++	+								
Cx. bicornutus				+++	+++		+	+		+							
Cx. cinctellus					+											+	
Cx. infantulus					++		++	++	+	+			++			++	
Cx. rubithoracis															+		
Cx. tuberis					+					+	+						
Cx. bitaeniorhynchus							+	+				+	++	+	+	++	
Cx. sinensis																+	
Lt. fuscana				+	+							++	+	+			
Lt. vorax			+	+	+		+	+				++	+	+		+	
Fi. ichiromiyagii						+										+	
Mi. elegans																++	
Mi. luzonensis															+	++	
Cq. crassipes																+	
Cq. ochracea																+	
Ma. uniformis																+	
Or. anopheloides			++	+++	+++												
Ml. genurostris	++																
To. yanbarensis		++															
Tp. yaeyamensis			++	+++	+++												
Tx. m. yaeyamae			++	++	++												
Tx. m. yamadai			++	++	++												
Tx. okinawensis				+	+												
Ur. jacksoni					+					++							
Ur. nivipleura			+														
Ur. n. ryukyuana			++	+++	+++			+									
Ur. ohamai							+		+	++	++						
Ur. tanakai						+											
Ur. yaeyamana										+							
Ur. annandalei					+		++	+	+								
Ur. lateralis													++			+	
Ur. macfarlanei							++	+	+	+							

Miyagi and Toma (2017). +, rare; ++, common; +++, abundant; ?, not clear.

付表 4　琉球列島産蚊の吸血源動物

種名	吸血源							吸血せず
	温血動物			冷血動物			不明	
	ヒト	哺乳類	鳥類	爬虫類	両生類	魚類		
モンナシハマダラカ	○	○						
オオツルハマダラカ	○	○						
ヤマトハマダラカ	○	○						
オオハマハマダラカ	○	○						
シナハマダラカ	○	○	○					
タテンハマダラカ	○	○						
コガタハマダラカ	○	○						
キンイロヤブカ	○	○	○	○	○			
オキナワヤブカ	○	○						
ヤエヤマヤブカ	○	○						
ニシカワヤブカ	○	○						
カニアナヤブカ					○	○		
ムネシロヤブカ	○	○						
アマミヤブカ	○	○						
サキシマヤブカ							○	
ナンヨウヤブカ	○	○						
ハマベヤブカ	○	○	○					
ワタセヤブカ	○	○						
ネッタイシマカ	○	○						
ヒトスジシマカ	○	○	○	○	○			
ダイトウシマカ	○	○						
ダウンスシマカ	○	○						
ミヤラシマカ	○	○		○				
リバースシマカ	○	○		○				
トウゴウヤブカ	○	○	○	○	○			
オオクロヤブカ	○	○	○	○				
アマミムナゲカ	○	○						
コガタフトオヤブカ	○	○		○				
アカフトオヤブカ	○	○		◎				
クロフトオヤブカ	○	○						
オビナシイエカ	○	○						
ジャクソンイエカ							○	
ミナミハマダライエカ	○	○						
シロハシイエカ	○	○	○					
ネッタイイエカ	○	○	○	◎	○			
ヨツホシイエカ	○	○	○					
コガタアカイエカ	○	○	○	○				
スジアシイエカ	○	○	○					
ニセシロハシイエカ	○	○	○					

付表4 （続き）

種名	吸血源							吸血せず
	温血動物			冷血動物			不明	
	ヒト	哺乳類	鳥類	爬虫類	両生類	魚類		
セシロイエカ	○	○						
クロフクシヒゲカ		○	○					
アカクシヒゲカ	○	○	○					
リュウキュウクシヒゲカ	○	○		○				
カギヒゲクロウスカ							○	
リュウキュウクロウスカ					○			
オキナワクロウスカ					○			
クロツノフサカ							○	
ハラオビツノフサカ		○	○					
フトシマツノフサカ		○	○	○	○			
アカツノフサカ					○			
カニアナツノフサカ							○	
カラツイエカ	○	○	○					
ミツホシイエカ	○	○						
サキジロカクイカ			○					
トラフカクイカ			○	○				
オキナワエセコブハシカ		○						
マダラコブハシカ					○			
ルソンコブハシカ		○			○			
ムラサキヌマカ	○	○	○					
キンイロヌマカ	○	○						
アシマダラヌマカ	○	○	○					
ハマダラナガスネカ			○					
オキナワカギカ								○
ヤンバルギンモンカ								○
ヤエヤマナガハシカ	○	○		○				
ヤエヤマオオカ								○
ヤマダオオカ								○
オキナワオオカ								○
カニアナチビカ				○	○			
ムネシロチビカ					○			
リュウキュウクロホシチビカ					○			
シロオビカニアナチビカ					○	○		
イリオモテチビカ							○	
ハラグロカニアナチビカ					○			
オキナワチビカ					○			
コガタチビカ		○			○	○		
マクファレンチビカ					○			

◎：初記録（當間・宮城，未発表）

Appendix 4. Blood-sucking sources of mosquitoes recorded from the Ryukyu Archipelago.

Species	Sources of bloodmeals							No blood-sucking
	Warm-blooded animals			Cold-blooded animals			Un-known	
	Human	Mammals	Birds	Reptiles	Amphibi-ans	Fishes		
An. bengalensis	○	○						
An. lesteri	○	○						
An. l. japonicus	○	○						
An. saperoi	○	○						
An. sinensis	○	○	○					
An. tessellatus	○	○						
An. yaeyamaensis	○	○						
Ae. v. nipponii	○	○	○	○	○			
Ae. o. okinawanus	○	○						
Ae. o. taiwanus	○	○						
Ae. nishikawai	○	○						
Ae. baisasi					○	○		
Ae. albocinctus	○	○						
Ae. j. amamiensis	○	○						
Ae. j. yaeyamensis							○	
Ae. lineatopennis	○	○						
Ae. vigilax	○	○	○					
Ae. watasei	○	○						
Ae. aegypti	○	○						
Ae. albopictus	○	○	○	○	○			
Ae. daitensis	○	○						
Ae. f. downsi	○	○						
Ae. f. miyarai	○	○		○				
Ae. riversi	○	○		○				
Ae. togoi	○	○	○	○	○			
Ar. subalbatus	○	○	○	○				
Hz. kana	○	○						
Ve. nobukonis	○	○		○				
Ve. atriisimilis	○	○		◎				
Ve. iriomotensis	○	○						
Cx. fuscocephala	○	○						
Cx. jacksoni							○	
Cx. mimeticus	○	○						
Cx. pseudovishnui	○	○	○					
Cx. quinquefasciatus	○	○	○	◎	○			
Cx. sitiens	○	○	○					
Cx. tritaeniorhynchus	○	○	○	○				
Cx. vagans	○	○	○					
Cx. vishnui	○	○	○					

Appendix 4 (continued).

Species	Sources of bloodmeals							No blood-sucking
	Warm-blooded animals			Cold-blooded animals			Un-known	
	Human	Mammals	Birds	Reptiles	Amphibi-ans	Fishes		
Cx. whitmorei	○	○						
Cx. nigropunctatus		○	○					
Cx. pallidothorax	○	○	○					
Cx. ryukyensis	○	○		○				
Cx. brevipalpis							○	
Cx. h. ryukyuanus					○			
Cx. okinawae					○			
Cx. bicornutus							○	
Cx. cinctellus		○	○					
Cx. infantulus		○	○	○	○			
Cx. rubithoracis					○			
Cx. tuberis							○	
Cx. bitaeniorhynchus	○	○	○					
Cx. sinensis	○	○						
Lt. fuscana			○					
Lt. vorax			○	○				
Fi. ichiromiyagii		○						
Mi. elegans					○			
Mi. luzonensis		○			○			
Cq. crassipes	○	○	○					
Cq. ochracea	○	○						
Ma. uniformis	○	○	○					
Or. anopheloides			○					
Ml. genurostris								○
To. yanbarensis								○
Tp. yaeyamensis	○	○		○				
Tx. m. yaeyamae								○
Tx. m. yamadai								○
Tx. okinawensis								○
Ur. jacksoni				○	○			
Ur. nivipleura					○			
Ur. n. ryukyuana					○			
Ur. ohamai					○	○		
Ur. tanakai							○	
Ur. yaeyamana					○			
Ur. annandalei					○			
Ur. lateralis		○			○	○		
Ur. macfarlanei					○			

◎ : New record (Toma and Miyagi, unpublished).

引用文献　References

Azar, D., Nel, A., Huang, D. and Engel, M. S. 2023. The earliest fossil mosquito. *Current Biology*, 33: 5240-5246.

Barraud, P. J., 1934. Diptera Family Culicidae, Tribe Megarhinini and Culicini. *Fauna Br. India*, 5: 1-463.

Bates, M. 1949. The Natural History of Mosquitoes. 378 pp., Mac-millan Co., New York.

Bohart R. M. and Ingram, R. L. 1946. Mosquitoes of Okinawa and Islands in the Central Pacific. *U. S. Navmed.*, 1055: 1-110, illus.

Borkent, A. and Grimaldi, D. A. 2004. The Earliest Fossil Mosquito (Diptera: Culicidae), in Mid-Cretaceous Burmese Amber. *Ann. Entomol. Soc. Am.*, 97: 882-888.

Bram, R. A. 1967. Contributions to the mosquito fauna of Southeast Asia (Diptera, Culicidae) II. The genus *Culex* in Thailand. *Contr. Am. Entomol. Inst.*, 2: 1-296, illus.

Clements, A. N. 1992. The biology of mosquitoes, Vol. 1. Development, Nutrition and Reproduction. 509 pp., Chapman and Hall, London.

Colless, D. H. 1959. Notes on the culicine mosquitoes of Singapore. VII. - Host preferences in relation to the transmission of disease. *Ann. Trop. Med. Parasitol.*, 53: 259-267.

Delfinado, M. D. 1966. The culicine mosquitoes of the Philippines, tribe Culicini (Diptera, Culicidae). *Mem. Am. Entomol. Inst.*, 7: 1-252, illus.

Ejiri, H., Sato, Y., Kim, K. S., Tamashiro, M., Tsuda, Y., Toma, T., Miyagi, I., Murata, K. and Yukawa, M. 2011. First record of avian *Plasmodium* DNA detection from mosquitoes collected in Yaeyama Archipelago, southwestern border of Japan. *J. Vet. Med. Sci.*, 73: 1521-1525.

Ejiri, H., Sato, Y., Sasaki, E., Sumiyama, D., Tsuda, Y., Sawabe, K., Matsui, S., Horie, S., Akatani, K., Takagi, M., Omori, S., Murata, K. and Yukawa, M. 2008. Detection of avian *Plasmodium* spp. DNA sequences from mosquitoes captured in Minami Daito Island of Japan. *J. Vet. Med. Sci.*, 70: 1205-1210.

Ejiri, H., Sato, Y., Sawai, R., Sasaki, E., Matsumoto, R., Ueda, M., Higa, Y., Tsuda, E., Omori, S. Murata, K. and Yukawa, M. 2009. Prevalence of avian malaria parasite in mosquitoes collected at a zoological garden in Japan. *Parasitol. Res.*, 105: 629-633.

福地斉志，當間孝子，宮城一郎．2021．「沖縄県民の森」における蚊の調査結果．沖縄県衛生環境研究所報，54：37-43.

Fukuchi, Y., Toma, T. and Miyagi, I. 2021. Mosquito survey in Okinawa Prefectural Forest, Okinawa, Japan. *Ann. Rep. Okinawa Pref. Inst. Health and Environment,* 54: 37-43.

Goma, L. K. H. 1966. The Mosquito. 144 pp., Hutchinson Tropical Monographs, London.

Higa, Y., Toma, T., Araki, Y., Onodera, I. and Miyagi, I. 2007. Seasonal changes in oviposition activity, hatching and embryonation rates of eggs of *Aedes albopictus* (Diptera: Culicidae) on three islands of the Ryukyu Archipelago, Japan. *Med. Entomol. Zool.*, 58: 1-10.

Ho, C. 1965. Studies on malaria in New China. *Chin. Med. J.*, 84: 491-497.

一盛和世（編）．2021．きっと誰かに教えたくなる蚊学入門　－知って遊んで闘って－．254

pp.，緑書房，東京.
池庄司敏明. 1993. 蚊 Mosquitoes. 246 pp.，東京大学出版会，東京.
Ikeshoji, T. 1993. The interface between mosquitoes and humans. 246 pp. University of Tokyo Press, Tokyo.
岩城操. 1989. ヌマカ属（アシマダラヌマカ）の幼虫および蛹の呼吸器の走査電子顕微鏡による形態観察. 日本熱帯医学雑誌，17：1-7.
上村清. 2016. 日本における蚊の分布と発生源. 衛生動物学の進歩　第2集（松岡裕之編），pp. 21-42. 三重大学出版会，三重.
Kamimura, K. 2016. Distribution and habitats of Japanese mosquitoes (Diptera: Culicidae). *In*: Progress of Medical Entomology and Zoology (From 1971 to 2015) (ed. Matsuoka H.), pp. 21-42, Mie University Press, Mie
上村清（編）. 2017. 蚊のはなし－病気との関わり－. 148 pp.，朝倉書店，東京.
Kim, K. S., Tsuda, Y. and Yamada, A. 2009. Bloodmeal identification and detection of avian malaria parasite from mosquitoes (Diptera: Culicidae) inhabiting coastal areas of Tokyo Bay, Japan. *J. Med. Entomol.*, 46: 1230-1234.
栗原毅. 2013. 日本の科学（2）近代化発足の頃の文献集. 198 pp.，有害生物研究会，神奈川.
Lorenz, C., Alves, J. M. P., Foster, P. G., Suesdek, L. and Sallum, M. A. M. 2021. Phylogeny and temporal diversification of mosquitoes (Diptera: Culicidae) with an emphasis on the Neotropical fauna. *Syst. Entomol.*, 46: 798-811.
Lukashevich, E. D. 2022. The oldest occurrence of Chaoboridae (Insecta: Diptera). *Russian Entomol. J.*, 31: 417-421.
Mannen, K., Toma, T., Minakawa, N., Higa, Y. and Miyagi, I. 2016. Biology of *Anopheles saperoi* Bohart and Ingram (Diptera: Culicidae), an endemic species in Okinawajima, the Ryukyu Archipelago, Japan. *J. Am. Mosq. Cont. Assoc.*, 32: 12-23.
宮城一郎. 1972. 実験室内での日本産蚊族の冷血動物吸血性について. 熱帯医学，14: 203-217.
Miyagi, I. 1972. Feeding habits of some Japanese mosquitoes on cold-blooded animals in laboratory. *Trop. Med.*, 14: 203-217.
Miyagi, I. 1976. Description of a new species of the genus *Topomyia* Leicester from the Ryukyu Islands, Japan (Diptera: Culicidae). *Trop. Med.*, 17: 201-210.
Miyagi, I. 1980. Notes on the Japanese species of the genus *Corethrella*, with the description of a new species (Diptera: Chaoboridae). *Jap. J. Sanit. Zool.*, 31: 15-21.
Miyagi, I. 1981. *Malaya leei* (Wharton) feeding on ants in Papua New Guinea (Diptera; Culicidae). *Jap. J. Sanit. Zool.*, 32: 332-333.
宮城一郎（編）. 2002. 蚊の不思議－多様性生物学. 254 pp.，東海大学出版会，東京.
Miyagi, I. (Ed.). 2002. Mosquitoes, their mysterious life. 254 pp. Tokai University Press, Kanagawa.
宮城一郎. 2023. ボウフラの採餌行動観察の面白さ，有用さ. ペストコントロール，204：60-62.
Miyagi, I. 2023. Fun and usefulness of observation on the feeding behavior of mosquito lar-

vae. *Pest Control,* 204: 60-62.

宮城一郎，当間孝子．1978．八重山群島の蚊科に関する研究．2．石垣島での 1975 年および 1976 年に採集したイエカ族の蚊について．衛生動物，29：305-312.

Miyagi, I. and Toma, T. 1978. Studies on the mosquitoes in the Yaeyama Islands, Japan. 2. Notes on the non-anopheline mosquitoes collected at Ishigakijima, 1975-1976. *Jap. J. Sanit. Zool.*, 305-312.

宮城一郎，当間孝子．1980．八重山群島の蚊科に関する研究．5．西表島の山脚，森林地帯で採集した蚊について．衛生動物，31：81-91.

Miyagi, I. and Toma, T. 1980. Studies on the mosquitoes in Yaeyama Islands, Japan. 5. Notes on the mosquitoes collected in forest areas of Iriomotejima. *Jap. J. Sanit. Zool.*, 31: 81-91.

宮城一郎，當間孝子．2017．琉球列島の蚊の自然史．217 pp.，東海大学出版会，神奈川.

Miyagi, I. and Toma, T. 2017. The natural history of mosquitoes in the Ryukyu Archipelago. 217 pp., Tokai University Press, Kanagawa.

Miyake, T., Aihara, N., Maeda, K. Shinzato, C., Koyanagi, R., Kobayashi, H. and Yamahira, K. 2019. Bloodmeal host identification with inferences to feeding habits of fish-fed mosquito, *Aedes baisasi*. *Scientific Reports*, 9: 4002. DOI: http://doi. nature.com/article/10.1038/s41598-019-40509-6

Miyata, A., Miyagi, I. and Tsukamoto, M. 1978. Haemoprotozoa detected from the cold-blooded animals in Ryukyu Islands. *Trop. Med.*, 20: 97-112.

宮田彬，宮城一郎，当間孝子．1987．琉球産冷血動物の住血性原生動物．沖縄生物学会誌，25: 21-34.

Miyata, A., Miyagi, I. and Toma, T. 1987. Detection of Haemoprotozoa from the cold-blooded animals collected in the Ryukyu Islands. *Biol. Mag. Okinawa*, 25: 21-34.

宮山　修．2020. オットンガエルの血を吸う蚊 . *Satsuma*, 166: 181-189.

Mogi, M. 1978. Intra- and Interspecific predation in filter feeding mosquito larvae. *Trop. Med.*, 20: 15-27.

Mogi, M. 1981. Studies on *Aedes togoi* (Diptera: Culicidae). *J. Med. Entomol.*, 18: 477-480.

茂木幹義．1999．ファイトテルマータ．213 pp.，海遊舎，東京.

Mogi, M. 1999. Phytotelmata: Small habitats support biodiversity. 213 pp., Kaiyusha Publishers Co. Ltd., Tokyo.

森章夫．1976．キンパラナガハシカ *Tripteroides bambusa* (Yamada) の無吸血産卵について．熱帯医学，17：177-179.

Mori, A. 1976. Autogeny in *Tripteroides bambusa* Yamada. *Trop. Med.*, 17: 177-179.

中村哲，宮城一郎，當間孝子．1987．具志頭村海浜における蚊相と発生水域について．沖縄県公衆衛生学会誌，18：104-108.

Okada, K. and Hara, J. 1962. Notes on a new species of the *Corethrell*a from Japan (Diptera: Chaoboridae). *Bull. Phys. Educ. Juntendo Univ.*, 5: 49-55.

Okazawa, T., Horio, M., Suzuki, H. and Mogi, M. 1986. Colonization and laboratory bionomics of *Topomyia yanbarensis* (Diptera: Culicide). *J. Med. Entomol.*, 23: 493-501.

Okudo, H., Toma, T., Sasaki, H., Higa, Y., Fujikawa, M. Miyagi, I. and Okazawa, T. 2004. A crab-hole mosquito, *Ochrelotatus baisasi*, feeding on mud-skipper (Gobiidae: Oxudercinae) in the Ryukyu Islands, Japan. *J. Am. Mosq. Cont. Assoc.*, 20: 134-137.

Reeves, L. E., Holderman, C. J., Blosser, E. M., Gillett-Kaufman, J. L., Kawahara, A. Y., Kaufman P. E. and Burkett-Cadena, N. D. 2018. Identification of *Uranotaenia sapphirina* as a specialist of annelids broadens known mosquito host use patterns. *Communications Biology,* 1: 92. DOI: https://doi.nature.com/commsbio/10.1038/s42003-018-0096-5

Reid, J. A. 1968. Anopheline mosquitoes of Malaya and Borneo. *Stud. Inst. Med. Res. Malaya*, 31: 1-520.

三條場千寿，比嘉由紀子，沢辺京子．2019．あなたは嫌いかもしれないけど，とってもおもしろい蚊の話．143 pp.，山と渓谷社，東京．

佐々学，栗原毅，上村清．1976．蚊の科学．312 pp.，図鑑の北隆館，東京．

Sasa, M., Kurihara, T, Kamimura, K. 1976. Mosquito science. 312 pp. Hokuryukan Co. Ltd., Tokyo.

Sawabe, K., Isawa, H., Hoshino, K., Sasaki, T., Roychoudhury, S., Higa, Y., Kasai, S., Tsuda, Y., Nishiumi, I., Hisai, N., Hamao, S. and Kobayashi, M. 2010. Host-feeding habits of *Culex pipiens* and *Aedes albopictus* (Diptera: Culicidae) collected at the urban and suburban residential areas of Japan. *J. Med. Entomol.*, 47: 442-450.

椎名誠．1998．蚊学ノ書．309 pp.，集英社文庫，東京．

Sirivanakarn, S. 1972. Contributions to the mosquito fauna of Southeast Asia. XIII. The genus *Culex*, subgenus *Eumelanomyia* Theobald in Southeast Asia and adjacent areas. *Cont. Am. Entomol. Inst.*, 8: 1-86, illus.

Sirivanakarn, S. 1976. Medical entomology studies - III. A revision of the subgenus *Culex* in the Oriental Region (Diptera: Culicidae). *Cont. Am. Entomol. Inst.*, 12: 1-272, illus.

砂原俊彦，比嘉由紀子．2016．蚊の行動研究における最近の展開．衛生動物学の進歩　第2集（松岡裕之編），pp.67-86．三重大学出版会，三重．

Sunahara, T. and Higa, Y. 2016. Recent progress in behavioral studies of mosquitoes. *In:* Progress of Medical Entomology and Zoology (From 1971 to 2015) (ed. Matsuoka H.), pp. 67-86, Mie University Press, Mie.

Tamashiro, M., Toma, T., Mannen, K., Higa, Y. and Miyagi, I. 2011. Bloodmeal identification and feeding habits of mosquitoes (Diptera: Culicidae) collected at five islands in the Ryukyu Archipelago, Japan. *Med. Entomol. Zool.*, 62: 53-70.

田中和夫．2014．Family Culicidae カ科．日本昆虫目録（日本昆虫目録編集委員会編集），181-201頁，櫂歌書房，福岡．

Tanaka, K. 2014. Family Culicidae. Catalogue of the insects of Japan (ed. Editorial Committee of Catalogue of the Insects of Japan), Touka Shobo, Fukuoka.

Tanaka K., Mizusawa, K. and Saugstad E. S. 1979. A revision of the adult and larval mosquitoes of Japan (including the Ryukyu Archipelago and the Ogasawara Islands) and Korea (Diptera: Culicidae). *Contrib. Am. Entomol. Inst.*, 16: 1-987.

Tanaka, K., Saugstad, E. S. and Mizusawa, K. 1975. Mosquitoes of the Ryukyu Archipelago. *Mosq. Syst.*, 7: 207-233.

Tempelis, C. H. 1975. Host-feeding patterns of mosquitoes, with a review of advances in analysis of blood meals by serology. *J. Med. Entomol.*, 11: 635-653.

羽鳥重郎．1919．八重山地方のマラリア．台湾医学会雑誌，205: 1054-1063.

Toma, T., Higa, Y., Okazawa, T. and Miyagi, I. 2007. Comparison of four light-trap methods for collecting mosquitoes in Iriomote Island, Ryukyu Archipelago, Japan. *Med. Entomol. Zool.*, 58: 1-10.

当間孝子，宮城一郎．1983．実験室内における *Tripteroides bambusa yaeyamensis* 抱卵蚊の色に対する反応．琉球大学医学雑誌，6: 96-103.

Toma, T. and Miyagi, I. 1983. Effect of color on oviposition of *Tripteroides bambus*a *yaeyamensis* in the laboratory. *Ryukyu Med. J.*, 6: 96-103.

Toma, T. and Miyagi, I. 1986. The mosquito fauna of the Ryukyu Archipelago with identification keys, pupal descriptions and notes on biology, medical importance and distribution. *Mosq. Syst.*, 18: 1-109.

Toma, T., Miyagi, I., Higa, Y., Okazawa, T. and Sasaki, H. 2005. Culicid and Chaoborid flies (Diptera: Culicidae and Chaoboridae) attracted to a CDC miniature frog call trap at Iriomote Island, the Ryukyu Archipelago, Japan. *Med. Entomol. Zool.*, 56: 65-67.

Toma, T., Miyagi, I. and Tamashiro, M. 2014. Blood meal identification and feeding habits of *Uranotaenia* species collected in the Ryukyu Archipelago. *J. Am. Mosq. Control Assoc.*, 30: 215-218.

Toma, T., Miyagi, I., Tamashiro, M., Higa, Y., Okudo, H. and Okazawa, T. 2011. Bionomics of the mud lobster-hole mosquito *Aedes* (*Geoskusea*) *baisasi* in the mangrove swamps of the Ryukyu Archipelago, Japan. *J. Am. Mosq. Control Assoc.*, 27: 207-216.

Toma, T., Tamashiro, M., Mizuta, H. and Miyagi, I. 2024. Blood-meal identification of seven species of mosquitoes (Diptera: Culicidae) collected in Japan. *Med. Entomol. Zool.*, 75: 13-15.

津田良夫．2013．蚊の観察と生態調査．359 pp.，北隆館，東京.

Tsuda, Y. 2013. Science watch: Field observations and ecological studies on Japanese mosquitoes. 359 pp., Hokuryukan Co. Ltd., Tokyo.

津田良夫．2017．鳥マラリアと媒介蚊に関する最近の研究．衛生動物，68：1-10.

Tsuda, Y. 2017. Review of recent studies on avian malaria parasites and their vector mosquitoes. *Med. Entomol. Zool.*, 68: 1-10.

津田良夫．2019．日本産蚊全種検索図鑑 . 127 pp., 北隆館，東京.

Tsuda, Y. 2019. An illustrated book of the mosquitoes of Japan: adult identification, geographic distribution and ecological note. 127 pp., Hokuryukan Co., Ltd., Tokyo.

Tsukamoto, M. and Miyata, A. 1978. Surveys on simian malaria parasites and their vector in Palawan Island, the Philippines. *Trop. Med.,* 20: 39-50.

Wilkerson, R. C., Linton, Y.-M. and Strickman, D. 2021. Mosquitoes of the World. Volume 1, 2. 1308 pp., Johns Hopkins University Press, Baltimore.

山田信一郎. 1932. 双翅目　蚊科. 日本昆虫図鑑(江崎悌三他), pp. 210-235.　北隆館, 東京.

おわりに

宮城は面白半分，暇つぶしに身近な蚊の行動を 8mmカメラやスマートフォンのファインダーを通して眺めていた．すでに承知済みと高を括っていたが、ボウフラの摂食行動や成虫の吸血，産卵行動はこれまでに理解していたのとは多くの点で異なり，目から鱗が落ちるような新たな感動を得た．この感動を独り占めするのでなく仲間と共有したいと，比較的時間に余裕がある定年退職された岡澤孝雄さん（金沢大学名誉教授)，當間孝子さん（元琉球大学教授)，水田英生さん（元検疫官)，それに現在，第一線で活躍中の比嘉由紀子さん（感染症研究所）にも動画を見る会に無理に参加していただいた． そのことにより情報や疑問はさらに多様化し，膨大な情報量が手元に蓄積された．また，動画の英語バージョンも作成してはとの意見に答えて，Yong Hoi-Sen 先生に英文校閲をしていただいき追加することができた．

小さな蚊の自然界での動画撮影には限度があり，狭い水槽や観察箱でクローズアップ撮影に頼ることになったので，自然界での行動とは多少異なるかもしれない．ある行動，たとえばヤンバルギンモンカの交尾行動が始まり完了するまでにはホバリング飛行が繰り返し観られ，数時間後にようやく交尾が完了することもある．長時間撮影した動画は編集に当たっては要所を切り貼りして 10 分以内に編集することになる．視聴者にはいかにもその行為が簡単に成立したかのように印象付けることになる．この点は動画の早送り操作や解説で補うことにした．編集作業は主として宮城と當間で行った．

2020 年初頭，本書が宮城・當間により企画されてから 5 年が過ぎようとしている．この間いろいろな予期せぬ不具合が生じた．コロナ禍で県内外への旅行ができなくなり，予定していた西表島での撮影ができなくなったこと．私たちには初めての動画編集ソフト「パワーダイレクター」を駆使するのに時間を要したこと．動画の編集（大容量）でパソコンが何度もパンクしたこと．高齢に伴いフィールド調査や長時間パソコンの使用ができなくなった．大家の都合により住み慣れた浦添の研究室を自宅へ移転することを余儀なくされたこともあったなど．

多くの先生，茂木幹義，都野典子，倉橋弘，葛西真治，三條場千寿，久野五郎に助言や励ましをいただいた．また，村山望（新星出版)，動物写真家の金城道夫，小松貴，村松稔，奥土晴夫，宮山修，万年耕輔さんには貴重な生態写真を提供していただいた．太田和鐘三さんには琉球列島に棲息する蚊を，神山さとしさんには蚊の採集の道具をそれぞれ描写していただいた．また，容量オーバーでパンクしたパソコンの修復や，編集ソフトの使い方などは斎藤育弘，我那覇隆伸さんにお世話になった．宮城敏夫（浦添総合病院)，宮城信行（天理市宮城医院)，徳西兵之助（天理市徳西医院）諸氏には私達の日頃の研究活動の意義を理解していただき物心両面から多大な支援をいただいた．あらためて諸氏にお礼を申し上げたい．このように多くの方の支援を得て本書がようやく出来上がった．出版に際して，有益な助言をいただいた池宮紀子，ボーダーインクの社長に謝意を表したい．

ここに収録した動画は沖縄に生息する一部の蚊の限られた行動に過ぎない．この動画を見ていただいた方々が各自の地域に生息する身近な蚊の生態を是非動画に収録していただきたい．この本がそのきっかけになったら幸いである．

Epilogue

In the beginning, I (Miyagi) was just interested in killing time and observing the behavior of mosquitoes breeding around my yard through the viewfinder of my 8 mm camera or smartphone. I had assumed I was already familiar with the feeding behavior of mosquito larvae and the blood-sucking and egg-laying behavior of adults. However, in many ways they were completely different from what I had previously understood. I was amazed and impressed, and the scales fell from my eyes.

Wanting to share this excitement with my colleagues, I asked retired people who have relatively more free time, including Dr. T. Okazawa (Professor Emeritus at Kanazawa University), Dr. T. Toma (Former Professor at the University of the Ryukyus), Dr. H. Mizuta (Former Quarantine Officer), and Dr. Y. Higa (National Institute of Infectious Disease), who is currently active on the front lines to join the video viewing endeavor. As a result, information and questions became more diverse, and a huge amount of information was accumulated at hand. Editing the videos was mainly done by Miyagi and Toma. In response to a request to create an English version of the video, we had the help of Dr. H.-S. Yong (Professor Emeritus of Genetics and Zoology, Universiti Malaya).

Video recording of small mosquitoes in the field was limited, and we relied on close-up photography in small aquariums or observation boxes, which might differ somewhat from their behaviors in the field. In some behaviors, such as the mating of *Topomyia yanbarensis*, repeated hovering flight was observed from the beginning to completion, with mating finally completed after several hours. When editing a long video, the cut-and-paste technique was used to reduce it to 10 minutes or less. This gives the viewer the impression that the act was easily accomplished. We therefore decided to compensate for this by fast-forwarding the video and providing captions.

Five years have passed since this book was planned by Miyagi and Toma. During this time, various unexpected problems have arisen. Due to the COVID-19 pandemic, we were no longer able to travel to record video within or outside the prefecture. It took us a while to get to grips with Power Director, a video editing software that was a new experience for us. Our computer crashed several times while editing videos (over capacity). Furthermore, due to unforeseen circumstances, we had to relocate our laboratory from Urasoe in Okinawa to Miyagi's house.

We would like to thank many scientists, including Dr. M. Mogi, Dr. N. Tsuno, Dr. H. Kurahashi, Dr. S. Kasai, Dr. C. Sanjoba, and Dr. G. Kuno for their advice and encouragement. Animal photographers, Mr. N. Murayama, Mr. M. Kinjo, Mr. T. Komatsu, Mr. M. Muramatsu, Mr. H. Okudo, Mr. O. Miyama, and Mr. K. Mannen provided us with invaluable photographs. Mr. S. Ohtawa and Mr. S. Kamiyama each provided illustrations of the mosquitoes that found in the Ryukyu Archipelago, and the tools used to mosquito collection.We would also like to thank Mr. I. Saito and Mr. T. Ganaha for their help in repairing our computer, which had crashed due to exceeding its capacity, and for helping us how to use the editing software. We would also like to thank Dr. T. Miyagi (Urasoe General Hospital), Dr. N. Miyagi (Miyagi Hospital, Tenri City), and Dr. H. Tokunishi (Tokunishi Hospital, Tenri City) for their understanding of the significance of our daily research activities and for their tremendous support, both materially and morally. We would like to express our gratitude to N. Ikemiya, President of Borderink Co. Ltd., for her helpful advice regarding the publication of this book.

This book was finally completed thanks to the support of so many people. The videos recorded here only show a limited number of the behaviors of some of the common mosquitoes that are found in the Ryukyu Archipelago. We hope that those who watch this video will also record videos of the behavior of familiar mosquitoes that are found in their own areas. We would be delighted if this book could inspire them to do so.

西表島後良川河口の夜明け
Dawn at the mouth of the Shiira River on Iriomotejima

索引 index

た

な

は

ま

や

ら

わ

A

B

著者紹介

宮城一郎 (みやぎ いちろう)
北海道大学大学院農学研究科博士課程修了
農学博士（北海道大学）
長崎大学熱帯医学研究所助手，琉球大学医学部教授を経て，現在，同大学名誉教授
琉球大学博物館（風樹館）協力研究員
専門は昆虫学，衛生動物学

當間孝子（とうま たかこ）
琉球大学農学部卒業，医学博士（鹿児島大学）
琉球大学医学部助手を経て，同大学医学部教授
現在，琉球大学博物館（風樹館）協力研究員
専門は衛生動物学

岡澤　孝雄（おかざわ たかお）
北海道大学大学院理学研究科博士課程動物学専攻
理学博士（北海道大学）
佐賀医科大学医学部助手，金沢大学医学部講師，留学生センター教授を経て，現在，同大学名誉教授，金沢大学国際感染症制御学 協力研究員
専門は衛生動物学

水田英生（みずた ひでお）
日本獣医畜産大学大学院獣医学研究科修士課程修了
獣医学修士（日本獣医畜産大学）
厚生省神戸検疫所専門職，厚生労働省関西空港検疫所衛生課長を経て，同省大阪検疫所次長，現在，検疫所衛生技官のカやノミなどの検疫害獣の研修の指導者
専門は衛生動物

比嘉由紀子（ひが ゆきこ）
長崎大学大学院医歯薬学総合研究科博士課程修了
医学博士（長崎大学）
長崎大学熱帯医学研究所助教を経て現在，国立感染症研究所昆虫医科学部分類生態室長
専門は衛生動物学

Authors

Ichiro Miyagi
Doctor Course, Graduate School of Agriculture, Hokkaido University, Ph. D. (Hokkaido University), Assistant Professor, Institute of Tropical Medicine, Nagasaki University, Professor, Faculty of Medicine, University of the Ryukyus, Professor Emeritus, University of the Ryukyus. Collaborative Researcher, University of the Ryukyus Museum (Fujukan).
Specialty Field: Entomology, Medical Zoology

Takako Toma
Graduate Faculty of Agriculture, University of the Ryukyus (BSc)
Ph. D. (Kagoshima University). Assistant Professor, Faculty of Medicine, University of the Ryukyus. Professor, Faculty of Medicine, University of the Ryukyus. Collaborative Researcher, University of the Ryukyus Museum (Fujukan) .
Specialty Field: Medical Zoology

Takao Okazawa
Doctor Course, Graduate School of Science, Hokkaido University , Ph. D. (Hokkaido University).
Assistant Professor, Saga Medical University
Lecturer, Kanazawa University. Professor, International Student Center, Kanazawa University.
Collaborative Researcher, Kanazawa University.
Specialty Field: Medical Zoology

Hideo Mizuta
Master Course, Graduate School of Veterinary Medicine, Nippon Veterinary and Animal Science University, M. Veterinary Medicine (Nippon Veterinary and Animal Science University) .
Specialist, Kobe Quarantine Station, Ministry of Health and Welfare, Chief, Health Division, Kansai Airport Quaratine Station, Ministry of Health, Labour and Welfare, Deputy Director, Osaka Quarantine Station, Ministry of Health, Labour and Welfare, Instructor, Training on quarantine pest animals for quarantine station sanitary technicians.
Specialty Field: Medical Zoology

Yukiko Higa
Doctor Course, Graduate School of School of Biomedical Sciences, Nagasaki University, Ph. D. (Nagasaki University).
Assistant Professor, Institute of Tropical Medicine, Nagasaki University.
Head, Taxonomy and Ecology Section, Department of Medical Insect Science, National Institute of Infectious Diseases.
Specialty Field: Medical Zoology

琉球列島の蚊
−動画で見る蚊の不思議な生態−
Mosquitoes in the Ryukyu Archipelago:
Amazing biology of mosquito on video

2025年3月31日　初版第1刷発行

著者　宮城一郎・當間孝子・岡澤孝雄・水田英生・比嘉由紀子著
I. Miyagi, T. Toma, T. Okazawa, H. Mizuta and Y. Higa

発行者　池宮紀子
発行所　ボーダーインク
〒902-0076　沖縄県那覇市与儀226-3
電話　098(835)2777　FAX 098(835)2840
https://www.borderink.com

印刷　株式会社東洋企画印刷

ISBN978-4-89982-478-7